N_2O and CH_4 Emission from Wastewater Collection and Treatment Systems

GWRC Report Series

N_2O and CH_4 Emission from Wastewater Collection and Treatment Systems

State of the Science Report and Technical Report

Edited By

Jeff Foley, Zhiguo Yuan, Jurg Keller, Elena Senante, Kartik Chandran, John Willis, Anup Shah, Mark van Loosdrecht and Ellen van Voorthuizen

Global Water Research Coalition

IWA PUBLISHING

Published by **IWA Publishing**
Alliance House
12 Caxton Street
London SW1H 0QS, UK
Telephone: +44 (0)20 7654 5500
Fax: +44 (0)20 7654 5555
Email: publications@iwap.co.uk
Web: www.iwapublishing.com

First publishing 2015

British Library Cataloguing in Publication Data
A CIP catalogue record for this book is available from the British Library

ISBN: 9781780407333 (Paperback)
ISBN: 9781780407340 (eBook)

Disclaimer

Contents

About the GWRC

Global cooperation for the exchange and generation of water knowledge

In 2002 twelve leading research organisations have established an international water research alliance: the Global Water Research Coalition (GWRC). GWRC is a non-profit organization that serves as a collaborative mechanism for water research. The benefits that the GWRC offers its members are water research information and knowledge. The Coalition focuses on water supply and wastewater issues and renewable water resources: the urban water cycle.

The members of the GWRC are:

KWR – Watercycle Research Institute (Netherlands), PUB – Public Utilities Board (Singapore), STOWA – Foundation for Applied Water Research (Netherlands), SUEZ Environnement – CIRSEE (France), TZW – German Water Center (Germany), UK Water Industry Research (UK), Veolia Environnement VERI (France), Water Environment Research Foundation (US), Water Quality Research Australia (Australia), Water Research Commission (South Africa), Water Research Foundation (USA), and the Water Services Association of Australia.

The US Environmental Protection Agency has been a formal partner of the GWRC since 2003. The Global Water Research Coalition is affiliated with the International Water Association (IWA).

GWRC members represents the interests and needs of 500 million consumers and has access to research programs with a cumulative annual budget of more than €150 million. The research portfolio of the GWRC members spans the entire urban water cycle and covers all aspects of resource management.

Preface

The Global Water Research Coalition is an international organisation that is dedicated to the exchange and generation of knowledge to support sustainable development and management of the urban water cycle. The research agenda is developed by the member organisations of the GWRC and reflects their priorities and recognises global trends and drivers that affect the urban water cycle. The present research agenda includes Climate Change as one of the priorities areas. This research area comprises topics related to the possible impact of climate change on the urban water sector as well as the possible contribution to climate change by the urban water sector via the direct and indirect emission of greenhouse gasses (GHG).

The objective of this joint effort was to collect and develop knowledge needed to understand and manage the emission of N_2O (nitrous oxide) and CH_4 (methane) by wastewater collection and treatment systems. Starting with a kick-off meeting in Vienna in September 2008, the GWRC members involved in this activity have bundled their individual research programs on this topic, aligned methodologies used and exchanged and discussed the resulting information of the programs and developed additional actions where needed. The outcomes were reviewed and discussed at a final workshop in Montreal in September 2010.

These activities has resulted in two reports: a State of the Science report which presents an overview of the current knowledge and know-how regarding the emissions of N_2O and CH_4 by wastewater collection and treatment systems and a Technical Report which includes all the details, facts and figures of the underlying studies used to develop the State of the Science report.

GWRC expresses the wish that our joint effort and resulting reports will be useful to all who are active in the field of understanding and control of greenhouse gas emissions by wastewater collection and treatment systems.

Frans Schulting

Managing Director GWRC

Acknowledgement

The Global Water Research Coalition wishes to express its appreciation to STOWA - Foundation for Applied Water Research (Netherlands) for acting as the GWRC's lead organisation for this joint effort and to recognise the high quality contributions by all organisations involved in this activity including Suez Environnement – CIRSEE (France), Water Environment Research Foundation (US), Water Research Commission (South Africa), and the Water Services Association of Australia. The support of the lead agent Stowa by Royal Haskoning is gratefully acknowledged as well.

The reports could not have been completed without the input and commitment of a number of individuals of the involved members of the GWRC and their associated organisations. These were:

Authors

Jeff Foley — GHD, Australia

Zhiguo Yuan, Jurg Keller — The University of Queensland, Australia

Elena Senante — CIRSEE-Suez, France

Kartik Chandran — Columbia University, USA

John Willis
Anup Shah

Brown and Caldwell
USA

Mark van Loosdrecht

Delft University of Technology
the Netherlands

Ellen van Voorthuizen

Royal HaskoningDHV

Contributors

Adam Lovell — WSAA, Australia

Lauren Fillmore — WERF, USA

Cora Uijterlinde — STOWA, the Netherlands

Gordon Wheale — UKWIR, UK

Pascal Dauthuille — CIRSEE-Suez France

Jo Burgess — WRC, South Africa

Summary

BACKGROUND

In a world where there is a growing awareness on the possible effects of human activities on climate change, there is a need to identify the emission of greenhouse gases (GHG) from wastewater treatment plants (WWTPs) (See Figure i). As a result of this growing awareness, some governments started to implement regulations that force water authorities to report their GHG emissions. With these developments, there exists a strong need for adequate insight into the emissions of N_2O (nitrous oxide) and CH_4 (methane), two important greenhouse gases. With this insight water authorities would be able to estimate and finally control their emissions. However, at this point few field data were available, with the result that the emission factors used by the Intergovernmental Panel on Climate Change (IPCC) were based on limited data. The lack of available data became the driver to start extensive research programs in Australia, France, the United States of America and the Netherlands with the objective to gain information needed to estimate, understand and control the emission of N_2O and CH_4 from wastewater collection and treatment systems.

Figure i Greenhouse gas emission from wastewater treatment plants.

CURRENT KNOWLEDGE

At the start of the research programs little was known about the processes which form N_2O,in contrast with the extensive knowledge on the formation of methane. In both cases, however, very little field data were available that gave insight on the level at which these two greenhouse gases were emitted from wastewater collection and treatment systems.

This lack of data resulted in the fact that the currently used IPCC emission factor for N_2O (3.2 g $N_2O \cdot person^{-1} \cdot year^{-1}$), which is used to estimate the N_2O emission from wastewater treatment plants, is based on only one field study in which the plant was not designed to remove nitrogen. Furthermore this lack of data has led the IPCC to conclude that: '*wastewater in closed underground sewers is not believed to be a significant source of methane*' (IPCC, 2006 a,b).

The data that has been published prior to the start of the research programs showed a very large variation in the level of N_2O emission. This is due to the fact from the fact that the formation of N_2O is a very complex process which can be performed by both nitrifying and denitrifying bacteria and is influenced by several process parameters. Denitrification in anoxic zones was in many cases indicated as the dominant source of N_2O emission from biological nitrogen removal processes.

JOINT EFFORTS

Since the topic of greenhouse gas emission from wastewater collection and treatment collection systems is of significance for the whole sector,the GWRC members[1] decided to join their individual research program results and support collaboration between their individual research partners. These joint efforts have led to an increased level of understanding on the processes forming N_2O emission from wastewater treatment facilities, the variety therein, and the contribution of methane emission from sewers and WWTPs. This increased level of understanding can already be used by the stakeholders of the GWRC members who are directly involved in the daily operation of wastewater collection and treatment systems.

Adjacent to the joint efforts of the GWRC members and individual research partners, the International Water Association (IWA) formed a Task group on the use of water quality and process models for minimising wastewater utility greenhouse gas footprints. The IWA Task Group is also collaborating with the GWRC researchers.

OBJECTIVES

The overall objectives of the different research programs were:

- Define the origin of N_2O emission.
- Understand the formation processes of N_2O.
- Identify the level of CH_4 emissions from wastewater collection and treatment systems.
- Evaluate the use of generic emission factors to estimate the emission of N_2O from individual plants.

---

[1]GWRC members were (in brackets the partner that performed the research): WERF, USA (Columbia University, Brown and Caldwell); WSAA, Australia (The University of Queensland); STOWA, the Netherlands (Delft University of Technology; Royal Haskoning)

BOUNDARIES

The main focus was to identify the level of emission, the variation therein and improve the knowledge of N_2O formation. Definition of mitigation strategies was outside the scope of most of the research as the knowledge on formation and orgin was too limited at the start of the research programs.

RESEARCH N_2O

Methodology

In all participating countries a wide range of WWTP types was selected with the expectation that differences between plant design and process conditions can help elucidate the factors influencing N_2O formation. The individual research partners used different methodologies (see Figure ii) to determine the emission of N_2O. The methodologies used in Australia, France, and the USA[2] were very suitable to gain insight in the formation processes of N_2O. The methodology used in the Netherlands, where the N_2O emission was measured in the total off-gas of covered WWTPs was very suitable to capture the variability of the emission. The use of different methodologies shows the complementary value of joint efforts to increase the level of knowledge on N_2O emission from WWTPs. For future work on this topic both methodologies will be required to finally estimate and control the emission of N_2O from WWTPs.

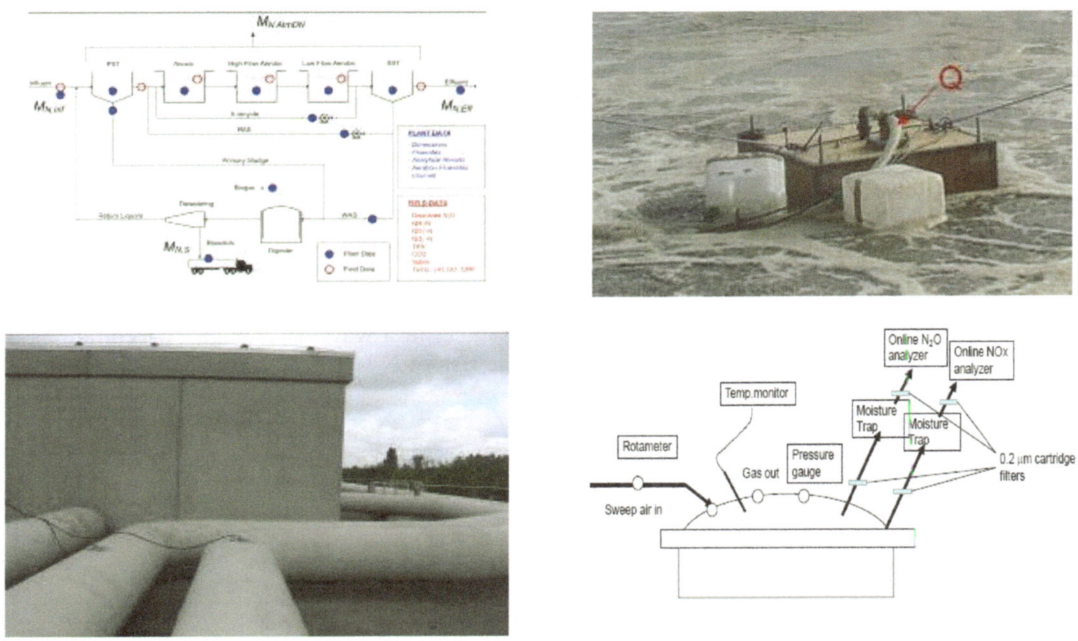

Figure ii Applied methodologies in the different research programs. Starting in the left corner above and then clockwise: Mass balance method based on liquid grab samples (Australia); Sampling box for aerated areas (France); Total off-gas measurements (the Netherlands); U.S. EPA, Surface emission isolation flux chamber (SEIFC); (USA).

[2]The protocol developed in the United States has been accepted by the USEPA, and is one of the most significant outputs of the research program.

Results

The emission of N_2O has been determined with different measurement protocols. For this reason it is not possible to average the emission numbers that have been derived. The results obtained in this research were suitable to increase the knowledge on N_2O formation and the variation therein, but the numbers can not be used to determine the emission from an individual plant as will be explained hereafter.
In line with earlier data, the field data in this study showed a large variety among the WWTP's sampled in the participating countries. The lowest emission that was measured was lower than 0.0001 kg N_2O-N/kg $TKN_{influent}$, while the highest reported emission was as high as 0.112 kg N_2O-N/kg $TKN_{influent}$. This lead to the following conclusions:

- The N_2O emission is highly variable among different WWTPs and at the same WWTP during different seasons or throughout the day.
- The use of a generic emission factor to estimate the emission from an individual WWTP is inadequate
- The emission from an individual WWTP can only be determined based on online measurements over the operational range of the WWTP (i.e. lowest temperature, highest load etc).

On the origin of the emission results showed that:

- The emission of N_2O mainly originates from nitrification, in contrast with earlier information.

At the start of the different research studies, very little was known about the process parameters that influenced the formation of N_2O, and most of the knowledge was based on laboratory studies. The joint efforts of the GWRC members and their research partners led to an increased level of understanding of the formation of N_2O and the process parameters influencing formation. It was concluded that:

- Nitrite accumulation leads to the formation of N_2O in aerobic zones as a result of low oxygen levels, sudden changes in ammonium load, and higher temperatures.
- High ammonium concentrations can lead to the emission of N_2O if nitrification occurs.

The above conclusions could already be translated to practice, in a way that if high concentrations of nitrite, ammonium or dissolved oxygen can be avoided the risk of N_2O emission can be reduced. It was concluded that:

- Systems that are not designed to remove nitrogen will have a high risk of N_2O emission if unintentional nitrification occurs.

With the present insight, it is possible to estimate the risk for N_2O emissions from a specific WWTP. This estimation can be based on the risk matrix presented in the following Table:

Risk on N_2O	**High risk**	**Medium risk**	**Low risk**
Parameter			
Effluent total organic nitrogen (mg/l)	>10	5–10	<5
Range in N-concentration in plant	H	M	L
Load variations (daily)	H	M	L
Maximum NO_2 concentration (mg N/l) anywhere in plant	>0.5*	0.2–0.5	0.2

*Risk does not increase at higher NO2 concentrations.

Based on the above matrix and the other conclusions the major conclusion of the research performed on N_2O emission from WWTPs is:

A good effluent quality (TN < 5 mgN/l) goes hand in hand with a low risk of N_2O emission

Remaining knowledge gaps and future research

Based on the outcomes of the research, valuable knowledge was gained to estimate and control the emission of N_2O from wastewater collection and treatment systems. The remaining knowledge gaps, their objectives and the type of research required are summarised as follows:

Knowledge gap	Objective	Future research
Insight in the variability of N_2O emission throughout the year at a WWTP to be able to define guidelines to design a sampling program at uncovered plants.	To obtain a good emission estimate of individual plants with minimal uncertainty.	Long term measurements in the total off-gas of WWTPs (covered ones are the most suitable to do so).
The relative contribution of autotrophic and heterotrophic processes to N_2O generation.	To develop mitigation strategies	High resolution monitoring of liquid phase N_2O specific zones of WWTP.
Mitigation strategies	To define measures to control emission via process design and control.	Measurements at different zones of one specific WWTP to study effect of different measures.
Emission from unknown sources like biofilm based processes and receiving aquatic environment.	To define level of N_2O emissions from these sources and to complete the picture of the whole urban watercycle.	Measurements at several locations that capture the variability that is expected.

RESEARCH CH_4

Methodology

The emission of methane was determined both from wastewater collection and treatment systems. The emission from wastewater collection systems was performed in Australia and the United States of America (see Figure iii). In Australia measurements were made in the liquid and gas phase in or around raising mains. The gas phase of unventilated lift stations was analysed in a study from the United States of America. A major obstacle in finally determining the emission of CH_4 (kg/d) from sewers is the determination of the gas flow (m^3/d). Developing a strategy for this obtaining flow measurement is one of the major research topics in this area. Mitigation strategies to control the emission of CH_4 from sewers were tested on laboratory and field level in Australia.

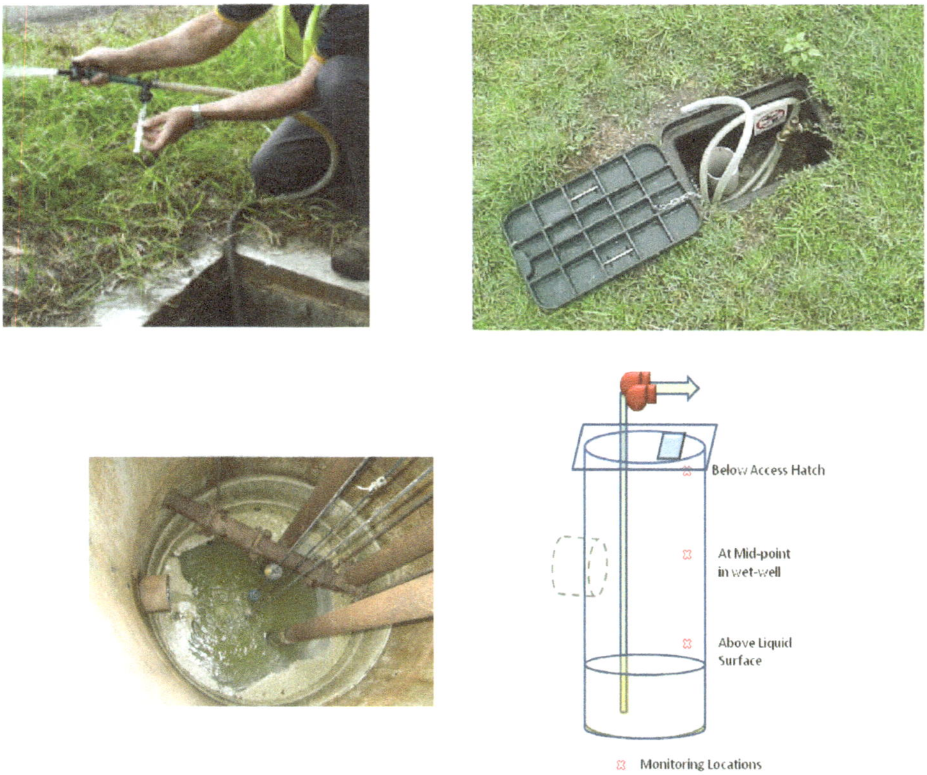

Figure iii Above: Sampling system rising mains (Australia); Under: Sampling system unventilated lift stations (USA).

The emission of CH_4 from wastewater treatment systems was investigated in France and the Netherlands. In France, the emission of CH_4 was monitored via a gas hood that was placed at the surface of different zones in a WWTP.

The emission of CH_4 in the Netherlands was determined based on grab samples taken from the different process units. These samples were taken in the same period as the emission of N_2O was monitored. In this way the carbon footprint of a WWTP could be determined as the data of electricity and natural gas use were readily available.

Results

At the start of the research, very little was known about the level of CH_4 emission from sewers and WWTP; the emission from sewers was even neglected. The results showed that the methane concentration in the liquid and gas phase from wastewater collection and treatment can be substantial. Concentrations up to more than 30 mg/l in the liquid phase were reported and emissions from lift stations were found to be as high as ~700 kg CH_4/year, but also emissions close to zero were found. This led to the following conclusion:

- Formation and emission from wastewater collection systems can be substantial and should not be neglected.

Measurements to define the emission of CH_4 (i.e. kg/d) from sewerage systems were found to be very difficult and complicated. Development of a good strategy measurement is seen as an important research topic.

Furthermore, a start was made to find strategies that could control the emission of CH_4 from sewers. Based on these preliminary experiments it was concluded that:

- Odour mitigation strategies in sewers likely also supports reduced CH_4 formation.

The level of CH_4 emission from WWTPs varied greatly from almost zero emission (< 0.0004 kg CH_4-COD/kg $COD_{influent}$) to emissions as high as 0.048 kg CH_4-COD/kg $COD_{influent}$). In general it was concluded that:

- Emission of CH_4 from WWTPs mainly originates from CH_4 formed in sewers and from sludge handling processes.

Remaining knowledge gaps and future research

Based on the outcomes of the research valuable knowledge was gained to estimate and control the emission CH_4 from wastewater collection and treatment systems. The knowledge gaps, their objectives and the type of research required are summarised as follows:

Knowledge gap	Objective	Future research
Strategy to determine amount of gas emitted to the air from wastewater collection systems.	To define the emission (kg/d) of CH_4 from wastewater collection systems	Develop a strategy based on field data.
Field data from different type of wastewater collection systems around the world.	To make a good estimate of the contribution of wastewater collection systems. To deliver data for the development, calibration and validation of CH_4 emission models.	Field measurements both liquid and gas phase from rising mains and gravity sewers around the world.
Cost effective mitigation strategies	To control the emission of CH_4 from wastewater collection systems.	Experiments in practice to study the effects and costs of different mitigation strategies.
Emission from sludge treatment lagoons.	To define level of CH_4 emissions from this source.	Measurements at several locations that capture the variability that is expected.

TOTAL CARBON FOOTPRINT

As a first indication on the possible contribution of N_2O and CH_4 emission to the total carbon footprint of a WWTP, the result in the Netherlands could be used as an example.

In the case studies in the Netherlands, the specific emissions of N_2O and CH_4 were determined at the same time. Together with the data on the related consumption of electricity and natural gas, it was possible to calculate a carbon footprint of three WWTPs. To determine the carbon footprint, all sources were converted to CO_2 equivalents.[3] The results in the Netherlands indicated that the emission of CH_4

[3]It should be noted that the conversion numbers are country specific and do depend on the used energy mix (i.e. brown coal versus wind or solar energy), which is of influence on the total carbon footprint of a WWTP.

and N_2O can significantly contribute to the total carbon footprint of a WWTP. This contribution can vary from 2% to almost 90% of the carbon footprint under extreme conditions for N_2O and 5 – 40% for CH_4. One should be aware that these numbers are specific for the Netherlands. In any other country, these numbers can differ greatly as there exist a great variation in the way wastewater and sludge is handled as well as the specific composition of the energy mix used. Furthermore these numbers can significantly differ depending on how the boundaries are set around the analysis. In case of the analysis performed for the three Dutch WWTPs the contribution of e.g. chemical use, and sludge incineration were not accounted for.

FUTURE ACTIVITIES

In the future the following activities will be developed by GWRC members and their researchers to further estimate and control the emission of GHG from wastewater collection and treatment systems:

- Long term measurements of both N_2O formation and process variablesfrom one WWTP, to gain insight in N_2O formation processes and the variability throughout the year.
- Mitigation strategies to gain insight in the possibilities to control the emission via process design and control.
- Development of a predictive model on N_2O production and emission.

Part I

State of the Science Report

Chapter 1
Introduction

1.1 BACKGROUND

In a world where there is a growing awareness of the possible effects of human activities on climate change, there is a need to identify the emission of greenhouse gases (GHG) from wastewater treatment plants (WWTPs).[1] As a result of this growing awareness, governments started to implement regulations that require water authorities to report their GHG emissions. With these developments there exists a strong need for adequate insight into the emissions of N_2O and CH_4. With this insight water authorities would be able to estimate and finally reduce their emissions. At the time little information was available on the formation of GHG, and the emission factors used by the IPCC are based on limited data. The limits of available data became the driver to start extensive field studies in Australia, France, the United States of America and the Netherlands with the objective to fill the knowledge gaps needed to estimate and reduce the emission of N_2O and CH_4 from wastewater collection and treatment systems.

The research programs were performed by partners[2] of the GWRC members WERF (United States of America), WSAA (Australia), CIRSEE-Suez (France) and STOWA (the Netherlands).

1.2 OBJECTIVES

The overall objectives of the different research programs[3] were:

- Define the origin of N_2O emission.
- Understand the formation processes of N_2O.

---

[1]The greenhouse gases associated with the activities at WWTPs are CO_2, CH_4 and N_2O. Of these gases, N_2O is the most important as it has a 300-fold stronger effect than CO_2. CH_4 is less strong than N_2O but still has a 25-fold stronger effect than CO_2. Nitrous oxide (N_2O) can be formed during the conversion of nitrogenous compounds in wastewater; methane may be emitted in the sewer system and during sludge handling. The emission of CO_2 from the biological treatment is part of short cycle (or biogenic) CO_2 and does not contribute to thecarbon footprint. However, some carbon in wastewater may originate from fossil fuel.

[2]Partners were: Columbia University, USA; Brown and Caldwell, USA; The University of Queensland, Australia; Delft University of Technology, the Netherlands, Royal Haskoning, the Netherlands.

[3]In the technical report (GWRC, 2011) that accompanies this State of the Art Report the objectives of the individual partners are mentioned.

- Identify the level of CH_4 emissions from wastewater collection and treatment systems.
- Evaluate the use of generic emission factors to estimate the emission of N_2O from individual plants.

1.3 BOUNDARIES REPORT

The research described in this report was the first extensive research on N_2O and CH_4 emission from wastewater collection and treatment systems. The main focus was to identify the level of emission, the variation therein and improve the knowledge on N_2O formation.

Definition of mitigation strategies was outside the scope of most of the research as the knowledge on formation and orgin was too limited at the start of the research. For methane some mitigation strategies were investigated and are reported here.

1.4 OUTLINE REPORT

The current state of knowledge prior to the start of the research is summarised briefly in chapter 2. A summary of the regulations that apply in countries participating in the GWRC report is presented in chapter 3. In chapter 4, the applied methodologies to determine the emission of CH_4 and N_2O are presented. The results of the different research are presented in chapter 5, after which the results are discussed in chapter 6. Finally the conclusions and recommendations for further research are presented in chapter 7. The details of all research programs are described in a technical report (GWRC, 2011) and the following reports of the individual GWRC members were used:

- WERF: Chandran, K., 2010, Greenhouse nitrogen emission from wastewater treatment operations, WERF report U4R07a.
- WSAA: Foley, J. and Lant, P., 2009, Direct Methane and Nitrous oxide emissions from full-scale wastewater treatment systems, Occasional paper No.24, Water Service Association of Australia.
- STOWA: Voorthuizen van, E. M., van Leusden, M., Visser, A., Kruit, J., Kampschreur, M., Dongen van, U. and Loosdrecht van, M., 2010, Emissies van broeikasgassen van rwzi (in Dutch, summary in English), STOWA report 2010–08.

Chapter 2

Current state of knowledge

2.1 N_2O FORMATION

Nitrous oxide can be produced during the conversion of nitrogen in WWTPs. Based on an extensive literature review; three processes have been identified as the main processes by which N_2O can be formed. These processes are presented in Figure 2.1. In addition, N_2O can be formed by chemical denitrification or during co-oxidation of ammonia to NO and N_2O by methanotrophic micro-organisms (Kampschreur, 2010). The main process parameters positively influencing the formation of N_2O are presented also in Figure 2.1. Those process parameters were found in several research papers and are summarized and explained in several reviews (Kampschreur et al., 2009, Foley and Lant, 2008).

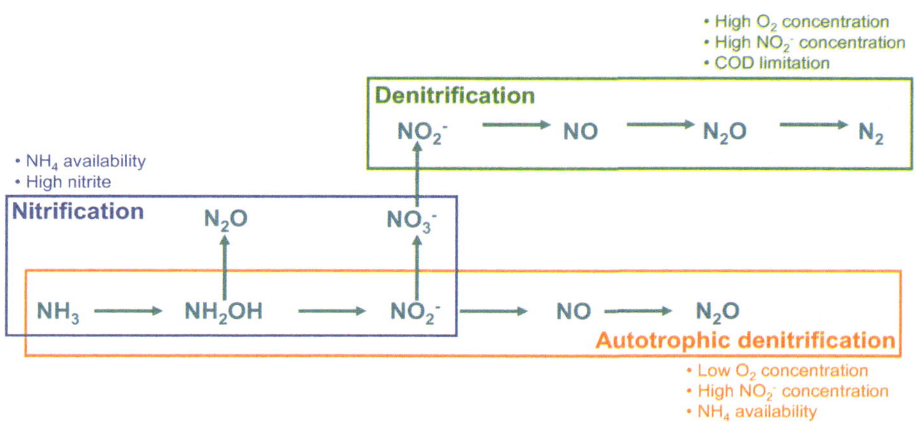

Figure 2.1 Schematic overview of the possible routes of N_2O emission and the process parameters that were found to influence the formation of N_2O.

Because of the different formation routes and the varying process parameters influencing those routes, a large variation in N_2O emission can be expected from full scale WWTPs. This is confirmed by the measurements that have been performed so far (Kampschreur et al., 2009). A variation between 0–15% of to the total nitrogen load of the WWTP was found to be emitted as N_2O (Kampschreur et al., 2009).

2.2 CH_4 EMISSION

In contrast to the formation of nitrous oxide, the formation of methane is very well known. However, little information is available on the amount of methane that is emitted from WWTPs and sewer systems.Prior to this research, only one study performed by Czepiel et al. (1993) measured the emission of methane from WWTPs. Data on emission of CH_4 from sewers were first reported by Guisasola et al. (2008). However, the possibility of methane formation in sewers has been acknowledged for a long time.

Methane can only be formed under anaerobic conditions. In this respect emission of methane can be expected from:

- sewer systems;
- influent works (formation in sewerage system, but emitted here);
- anaerobic/anoxic tanks as part of activated sludge systems;
- sludge digestion and handling.

Methane that is emitted from the influent works is formed in the sewer system. Little information is known about the amount of methane formed in sewers. This is in contrast with H_2S. In most cases the influent headworks is totally covered and emission of methane occurs after air treatment. Due to the anaerobic conditions in anaerobic and anoxic tanks methane might be formed. However, if there is a presence of aerobic zones methanogens will not survive.

At WWTPs that are equipped with an anaerobic sludge digester, methane can be emitted from different locations related to the digester. Methane can be emitted during the different processes of biogas combustion (leakages, incomplete combustion) or during storage of digested sludge. At WWTPs without sludge digesters methane formation is possible if the excess sludge is stored. The amount of methane emitted from these storage facilities will depend on the sludge retention time applied in the activated sludge system, the temperature and the level of dissolved methane, which in turn depends on the type of transport system prior to the WWTP.

Despite the presence of oxygen in the aeration tanks, methane can be emitted from these tanks. This is most likely methane that has been formed earlier in the process or in the sewer. The presence of methanogens in activated sludge has been proven by different authors (Lens et al., 1995 and Gray et al., 2002). In these studies it was proven that the contribution of the methane production by methanogens was very limited. Gray et al. (2002) reported 0.01–0.02% of the amount of carbon removed.

2.3 EMISSION FACTORS

The United Nations Framework Convention on Climate Change is the globally recognised basis for collective action on the reduction of anthropogenic greenhouse gas emissions (UNFCCC, 2007). One of the key obligations for signatory countries under the UNFCCC is the compilation of an annual national greenhouse gas (GHG) inventory, covering four general sectors (energy; industrial processes; agriculture, forestry and other land use; and waste). Emissions of methane and nitrous oxide from wastewater treatment and discharge are reported under the waste sector (IPCC, 2006b). However, GHG emissions are not usually measured directly, but estimated through the application of models that link emissions to data on observable activities (Foley and Lant, 2009).

2.3.1 Nitrous oxide (N_2O)

In the revised 1996 IPCC Guidelines the estimation methodology for N_2O emission from wastewater handling assumed minimal nitrogen removal during treatment, and hence all influent nitrogen is discharged to aquatic receiving environments where the nitrogen is converted. During this conversion, some of the discharged nitrogen will be emitted to the atmosphere as N_2O at a default factor of 0.01 kgN_2O-N/kgN discharged.

This value was revised in 2006 to 0.005 $kgN_2O\text{-}N/kgN^{-1}$discharged (IPCC, 2006a). In the same revision IPCC acknowledged that in many advanced WWTPs nitrogen removal occurs. The proposed default emission factor was 0.0032 kgN_2O[1]$\cdot person^{-1}\cdot yr^{-1}$, based on one full-scale study by Czepiel et al. (1995) on a basic secondary treatment plant without nitrogen removal.

The above mentioned IPCC guidelines are used by most countries to estimate the emission of N_2O from domestic WWTP for their national inventory reports (NIR). Andrews et al. (2009) examined ten country-specific NIRs, six countries used the IPCC default procedures, and four countries use their own factors.

2.3.2 Methane (CH_4)

In the current IPCC guidelines only methane emission from wastewater treatment systems is considered, the contribution of methane from sewers is neglected as can be read in the IPCC guidelines: '*wastewater in closed underground sewers is not believed to be a significant source of methane*' (IPCC, 2006a).

For wastewater treatment systems a generalised approach is prescribed (IPCC, 2006a). The exact description can be found in the IPCC documents (IPCC, 2006a). In essence, the approach is a reconciliation of the estimated mass of methane produced in the treatment process, with the measured mass of methane captured in the associated biogas system. Any difference in these figures is assumed to be a loss of methane to the atmosphere.

The above mentioned procedure is used by most countries to estimate the emission of CH_4 from WWTPs for their national inventory reports (NIR). Andrews et al. (2009) examined 12 country-specific NIRs, eight countries used the IPCC default procedures.

[1]Assuming a wastewater nitrogen loading of 16 $g\cdot person^{-1}\cdot d^{-1}$ for developed countries (i.e. high protein intake) (Tchobanoglous et al., 2003; IPCC, 2006a; DCC, 2008b), this equates to approximately 0.035% of the nitrogen load of the influent.

Chapter 3
Regulations around greenhouse gas emissions

As stated in the introduction, governments around the world started to implement regulations on reporting and reducing the emission of greenhouse gases. A summary of the regulations in the countries, of which the research is described here, is presented below.

The following Act and Regulations define the legislative requirements for greenhouse gas reporting within **Australia** which also counts for the wastewater treatment sector:

- *The National Greenhouse and Energy Reporting Act 2007*, including amendments; and
- *The National Greenhouse and Energy Reporting Regulations 2008*, including amendments:

The Regulations provide detailed requirements for reporting under the Act, including definitions of operational control, facilities, the requirements for registration and the types of greenhouse gas emissions and energy consumption/production that have to be reported. More details can be found in the GWRC technical report (GWRC, 2011).

In the **United States of America** some regulations on the emission of GHG can affect the wastewater treatment sector in the future. Those regulations fall under the attainment New Source Review program of the federal Clean Air Act (CAA), the Prevention of Significant Deterioration (PSD) program, and the federal mandatory reporting rule, separate from the CAA, that applies in general to those stationary sources that emit more 25,000 MT CO_2e per year. The California Global Warming Solutions Act of 2006 affects the wastewater sector in California only. This Act establishes the first comprehensive GHG regulatory program in the United States, and commits California to achieving significant GHG emission reductions by 2020. With this act five regulatory measures are already directed at the water/wastewater sector including increasing water use efficiency, increasing water recycling, reduction in the magnitude and intensity of energy use in California's water systems, increased usage of urban runoff, increased renewable energy production from water systems and a public goods charge on water meters to pay for most of the above activities.

In **France** there are no specific regulations concerning GHG emissions for the water sector. Nevertheless, all industries/companies with more than 50 employees must perform a carbon footprint assessment (new law July 2010). In the case of wastewater treatment plants, the French EPA (ADEME) recommends taking into account direct emissions of CH_4 and N_2O.

Treatment of wastewater in the **Netherlands** is delegated to the Waterboard Authorities. These Waterboard Authorities are obliged to report the emission of greenhouse gases from wastewater treatment plants with a capacity higher than 136,360 p.e[1] or from wastewater treatment plants that handle more than 50 tonne sludge per day (IPCC).

Since April 2010 the Dutch Waterboards signed a 'Dutch Climate Agreement' with the government. Part of this agreement is that the waterboards committed themselves to reduce the emission of N_2O and CH_4 with 30% (equal to approximately 200 ktonne CO_2-equivalents from 1990 to 2020).

[1]Based on 136 g total oxygen demand.

Chapter 4
Methodology

4.1 FIELD SAMPLING SITES N_2O

In total 26 WWTPs were monitored among the four countries participating in this research, namely, Australia, France, United States of America and the Netherlands. In all countries, a wide range of WWTP types was selected with the expectation that differences between plant design and processes conditions can help elucidate the factors influencing N_2O formation. Furthermore differences in climate conditions were taken into account for the selection of WWTPs in the USA and Australia. With respect to nitrogen and phosphorus removal, nine WWTPs removed phosphorus biologically, 21 were BNR systems and 5 of them were non BNR systems. The configurations that were present among the monitored WWTPs were: oxidation ditches (2), plug flow reactors (9), carrousel (5), SBR (1), MBR (1), and unknown (8). The characteristics of all WWTPs are summarised in Appendix 1.4.

4.2 SAMPLE COLLECTION AND ANALYSIS N_2O

4.2.1 Sample collection

Most WWTPs worldwide are not covered; therefore methods needed to be developed to determine the emission of N_2O. For uncovered WWTPs, the samples for the analyses of N_2O were collected via suspended gas hoods in the process basins. At WWTPs that were totally covered, samples could be collected from the total off-gas. More details of the sample collection in the different countries are summarised in Table 4.1.

In France and the USA, the gas samples were collected via a suspended (floating) gas hood. The working principles of these gas hoods are presented in Appendix 1.1.

Table 4.1 Overview of sample collection for N_2O measurements.

Country	Sample collection	Period	Time	Location in WWTP	Liquid/gas phase	Gas hood/ Total off gas
Australia	Grab samples	Winter/Spring 2008; Ideally 4 rounds.	Per round: 2–4 hours per day for 2 days	Anaerobic, anoxic, aerobic zones	Liquid	-
France	Continuously	*Information not available*	*Information not available*	Anaerobic, anoxic, aerobic zones	Gas	Gas hood
USA	Continuously/Grab samples	Summer/early Fall & Winter/early Spring	1 day (1/min; gas phase) 4–5x day (liquid)	Anaerobic, anoxic, aerobic zones Geometric center	Liquid (grab) & Gas (grab and continuous)	Gas hood
Netherlands	Continuously	Fall/Spring/Winter	7 days	-	Gas	Total off gas

4.2.2 Concentration

The methods used to measure the concentration of N_2O in either the gas or liquid phase are summarised in Table 4.2.

Table 4.2 Overview used methods to measure N_2O concentration in either gas or liquid phase.

Country	N_2O gas phase	N_2O liquid phase
Australia	-	Clark-type microsensor (N_2O 25 with 70 μm outside tip diameter), Unisense A/S, Aarhus, Denmark
France	Servomex model 4210 gas analyser	Via liquid phase sampling*
USA	Infrared	Miniaturised Clark-type sensor with internal reference and guard cathode; Unisense A/S, Aarhus, Denmark
Netherlands	Infrared; Emerson Process management Rosemount Analytical	-

*Details of the analysis in Appendix 1.2.

4.2.3 Gas flow

To determine the emission of N_2O, an accurate measurement of the gas flow is important. It is more complicated to measure the emission of N_2O from uncovered WWTPs than from WWTPs that are totally covered. The methods used at both covered and uncovered WWTPs are summarised below. More details on the gas hood as used in France and the United States of America can be found in Appendix 1.2.

France

The resulting areal flow of gas measured is calculated from the concentration measured at the exit of the sampling system (N2O or CH4 in mg/m^3 air) and from the air flow applied to the sampling system (m^3/h). The flow is related to the sampling area (area of the *flow chamber* in the case of non-aerated surfaces or the area of the *sampling box* in case of aerated surfaces), and is expressed in mg.h^{-1}.m^{-2}. (For the working principle of the flow chamber and sampling box see Appendix 1.1).

United States of America

Sampling procedures-measurement of advective gas flow rate from aerated zones

Advective flow of gas through the flux-chamber ($Q_{emission}$) in aerated zones was measured using a modification of ASTM method D1946. Briefly, a tracer gas consisting of 100,000 ppmv ($C_{helium-tracer}$) He was introduced into the flux-chamber at a known flow rate, Q_{tracer} (equation 1). He concentrations in the off-gas from the flux-chamber ($C_{helium-FC}$) were measured using a field gas-chromatograph equipped with a thermal conductivity detector (GC-TCD). $Q_{emission}$ was computed using equation 1.

$$Q_{tracer} * C_{helium-tracer} = (Q_{tracer} + Q_{emission}) * C_{helium-FC}$$

$$Q_{emission} = \frac{Q_{tracer} * (C_{helium-tracer} - C_{helium-FC})}{C_{helium-FC}} \qquad (1)$$

Sampling procedures-measurement of advective gas flow rate from non aerated zones

The only modification to the protocol to measure the emission flow rate from *non-aerated zones* was the introduction of sweep gas (air) or carrier gas through the flux-chamber at a known flow rate (Q_{sweep}), in addition to the He tracer gas. The corresponding $Q_{emission}$ was computed using equation 2. Addition of sweep gas is needed to promote mixing of the SEIFC (Surface emission isolation flux chamber) contents, owing to the low advective gas flow from the anoxic-zone headspace. Sweep-air N_2O concentrations were always measured and typically below the detection limits of the N_2O analyzer.

$$Q_{tracer} * C_{helium-tracer} = (Q_{tracer} + Q_{sweep} + Q_{emission}) * C_{helium-FC}$$

$$Q_{emission} = \frac{Q_{tracer} * (C_{helium-tracer} - C_{helium-FC})}{C_{helium-FC}} - Q_{sweep} \qquad (2)$$

During continuous N_2O measurements, $Q_{emission}$ was determined several times a day.

The Netherlands

The flow of off gas through the pipes was measured using a pitot tube. The pitot tube was connected to a pressure probe (Testo 445, max. 10 hPa, accuracy +/- 0.03 hPa), which transferred the pressure difference to a gas velocity. In this way the gas velocity was measured at least three times during the measuring campaign. During each measurement the gas velocity was measured at 20–50 individual points throughout the whole pipe. Based on the internal diameter of the pipe the gas flow was calculated. The pitot tube was used during the measurement campaigns in Papendrecht and Kralingseveer. During the measuring campaign at Kortenoord, the gas velocity was measured with a hot wire anemometer (Testo 435-1, max. 20 m/s, accuracy +/– 0.03 m/s and +4% of measuring value). After the measurements, the error of the measurement was determined. If this error was larger than the variation in the gasflow, the emission was calculated with an average gasflow, if the error was smaller than the emission was calculated with the daily gasflow.

4.2.4 Additional data

During all measuring campaigns, additional data which could be related to the emission of N_2O were collected to monitor the performance of the plant.

The additional data that were collected are:

- WWTP characteristics (process elements, volumes etc.);
- Influent, effluent composition and sludge concentration;
- On-line data from the available sensors at the WWTP.

Detailed information about the exact analyses and collection methods for the additional data can be found in the different research reports (Foley et al., 2009, Chandran, 2010; van Voorthuizen et al., 2010).

4.2.5 Quality control

The analysers used in the different research (Australia and the Netherlands) were calibrated prior to every measuring campaign. Details about the calibration methods can be found in the different research reports (Foley et al., 2009; Chandran, 2010; van Voorthuizen et al., 2010). The protocol that was developed for the measurements in the USA was validated in a special program in which the protocol was compared with other measuring methods. Details about this validation can be found in the different research reports (Foley et al., 2009;, Chandran, 2010; van Voorthuizen et al., 2010).

4.2.6 Calculations

The emission of N_2O was in general calculated using the following formula:

$$\text{Emission} = \frac{[N_2O] \cdot Q_{gas} \cdot 24}{Q_{influent} \cdot [TKN]} \tag{3}$$

In which:

- $[N_2O]$ is the concentration in the gas phase, either expressed as g N_2O or N_2O-N/m^3
- Q_{gas} is the gas flow expressed as m^3/h
- $Q_{influent}$ is the influent flow to the WWTP expressed as m^3/d
- [TKN] is the concentration NKj in the influent expressed as gN/m^3

In Australia the concentration of N_2O in the gas phase was determined based on the liquid phase measurements and k_La values. These k_La values were determined at laboratory scale and were corrected for the circumstances at full scale plants. More details about this methodology can be found in the research report (Foley, J. et al., 2009)

In the United States the surface flux calculated from the flux chamber was translated into the flux of a given zone by multiplying over the specific zone area.

4.3 FIELD SAMPLING SITES CH_4

4.3.1 Sewers

Australia

Two measurement campaigns have been performed on the emission of methane from sewers. One measurement campaign was performed in Australia, where liquid phase measurements were performed in two rising mains. In this report the result of one of the rising mains will be presented. This rising main (CO16[1]) is situated at the Gold Coast, Queensland, Australia (see Figure A1.5.1 in Appendix 1.5). Samples were taken at four sampling points (Pumping Station, Sample Points 1, 2 and 3) and sampling was repeated four times at an interval of one hour. In this way it was possible to capture wastewater samples with HRT in the sewer line for 0–8.7 hours.

United States of America

In the USA, gas phase measurements were performed in 64 lift stations on a force main system. The lift stations were predominantly fed by gravity sewers and in more limited cases by rising mains or a combination of the two.

The lift stations are situated in DeKalb County, Georgia, from which 60 are small with capacity ranging from 80 to 700 gallons per minute (gpm) (0.3–2.6 m^3/minute) and 4 are large with firm capacities of over 2,000 gpm (~7.6 m^3/min). Only one of the 64 lift stations is ventilated the other 63 are unventilated. An overview of DeKalb County and the locations of the 64 lift stations is presented in Appendix 1.3. The measurements were performed in a both cold (31 March–4 April 2009) and warm (13–17 July 2009) periods.

4.3.2 WWTPs

The emission of methane from WWTPs was investigated in France and the Netherlands. For both countries counts that the WWTPs investigated were the same as for the N_2O measurements (see section 4.1)

4.4 SAMPLE COLLECTION AND ANALYSIS CH_4

4.4.1 Liquid phase sample collection from sewers and analysis (Australia)

Wastewater sampling from rising mains was done through a special sampling arrangement (Figure 4.1). It consists of a 16 mm diameter pipe connecting a sampling tap at the ground level to the tapping arrangement attached to the underground pipe.

---

[1]The CO16 rising main receives domestic wastewater. It has an internal pipe diameter of 300 mm, giving an A/V ratio of 13.3 m^{-1}. The average daily wastewater flow was approximately 700 m^3. The pumping station was operated intermittently with 30 -40 pumping events per day, each lasting for 4–6 min in duration. During each pumping event, the calculated flow velocity was approximately 0.90–1.01 m/s.

Samples were collected using a hypodermic needle and 5 ml plastic syringe, attached directly to the pressurized rising main via a flexible hose. This procedure avoided any contact of the wastewater with atmosphere and possible oxygen interference.

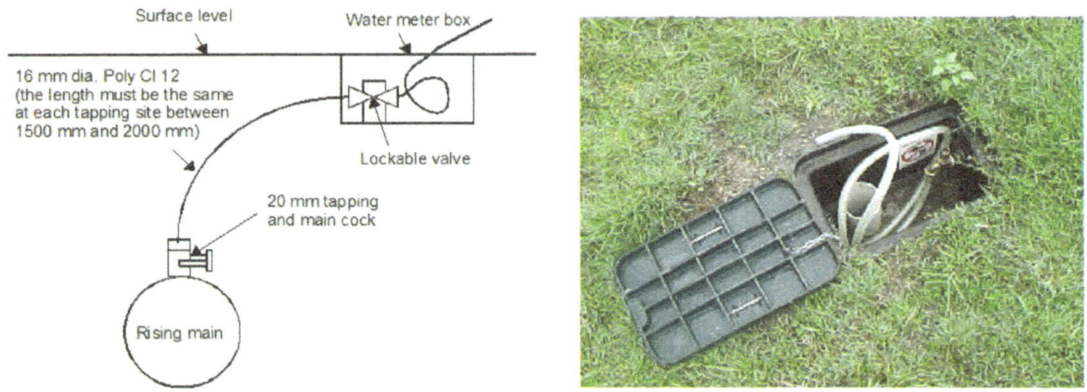

Figure 4.1 Sampling system for rising mains to determine concentration of CH_4 in liquid phase.

A sample collected was subsequently injected into freshly vacuumed BD Vacutainer® tubes through a 0.22 μm pore diameter unit (Millipore, Millex GP). The Vacutainer tube was mixed overnight in a shaker to allow equilibration of gas and liquid phases. Most of the methane (~97% at 25°C) would be transferred to the gas phase in this process (Alberto et al., 2000). The methane concentration in the gas phase of the tube was measured using a Shimadzu GC-9A Gas Chromatograph equipped with a flame ionization detector (FID). The concentration of methane in the initial liquid phase was then calculated using mass balance and Henry's law (Guisasola et al., 2008). The Vacutainer tube was weighed before and after sampling to determine the sample volume collected. This volume, along with the known volume of the Vacutainer tube, enables to calculate dissolved methane contained in the original wastewater sample.

4.4.2 Sample collection and analysis gas phase sewers (USA)

Sample collection in unventilated lift stations

Figure 4.2 shows the inside of a typical unventilated lift station in the DeKalb collection system. CH_4 concentrations were measured at three locations in each wet well during each sampling event: immediately above the liquid surface, mid-way up the wet well headspace; and immediately below the access hatch. While all three concentrations were collected, the highest of the three was used in order to somewhat mitigate the lack of accounting for dispersion of lighter-than-air CH_4 in between pumping cycles or the air dilution that results when air is pulled into the well during a pumping cycle. A similar procedure for data collection was used at large unventilated lift stations with open fore bays including LCC PS-3 and Honey Creek.

Figure 4.2 Sampling method used at unventilated lift stations.

Sample collection in ventilated lift stations

There is only one ventilated lift station within DeKalb's collection system. At LCC-1 the wetwell is covered and foul air exhaust fans continuously withdraw air and send it to an odor control system. Openings for the influent screens and at the wetwell covers allow air to enter the wetwell as foul air is exhausted. Instantaneous measurements for CH_4 and other parameters were sampled directly from the discharge header of the exhaust fan upstream of the biofilter.

Analysis

Portable instruments were used for taking instantaneous readings in the field. A hand-held flame ionization detector (FID, MicroFID manufactured by PhotoVac) with hydrogen as a fuel source was used to measure CH_4. A portable four-gas analyzer with photoionization detector (PID, RKI Eagle Gas Portable Monitor) was used to measure H_2S, CO_2, %lower explosive limit (LEL) and CO. Ambient air and raw sewage temperatures were also monitored to check for a correlation with the measured CH_4 concentration. A dissolved oxygen (DO) probe (YSI 550A DO) was used to measure ambient air temperature, raw sewage temperature, and dissolved oxygen concentration in the raw sewage.

4.4.3 Sample collection and analysis at WWTP

France

The measurement of methane at the WWTPs in France was performed with the same set-up as for the N_2O measurements as described in 4.2.1.

The Netherlands

Grab samples for the analysis of the methane concentration were taken at the major process units of the WWTP. The location of and the amount of grab samples taken at each WWTP are summarized in Table 4.3.

The grab samples were analyzed for CH_4 at the laboratory using gas chromatography (Varian CP-3800 Gas Chromatograph). The gas flow was determined in the same way as for the emission of N_2O (see section 4.2.3).

Table 4.3 Summary locations methane measurements and number of samples taken.

WWTP	Sample location	Number of samples	
Papendrecht	• After inlet work and coarse screen	3	
	• Selector	2	
	• Anaerobic tank	2	
	• Sludge loading	1	
	• Carrousels	1	
Kortenoord	• After inlet work and coarse screen and grid removal	5	
	• Selector	5	
	• Aeration tank (3 channels N_2O measurements)	3	
	• Sludge thickeners	3	
	• Sludge storage	3	
	• SludgVe dewatering site	4	
Kralingseveer		October	February
	• After inlet work and coarse screen	2	3
	• Primary settling	2	3
	• Selector	2	3
	• Sludge thickeners	2	3
	• Sludge storage	2	3
	• Aeration tank	2	2
	• Carrousel 1 and 2	2	3

The emission of CH_4 is based on the following measurements:

- CH_4 concentration in grab samples;
- Gas flow;
- Influent flow;
- COD concentration influent (after primary settling, including internal flow).

The CH_4 emission was calculated based on the average concentration in the grab samples. In almost all cases[2] the average concentration was multiplied by the average gas flow as the error in the measurements was larger than the variations in the gas flow.

[2]Exceptions were selector Papendrecht and sludge storage in Kralingseveer (February), here the daily gas flow was used.

4.5 MITIGATION STRATEGIES SEWERS

Chemical dosage or pH elevation to sewage is commonly used for the control of sulphide formation in sewers, or its transfer from wastewater to sewer air.

The effect of pH elevation and the dosage of nitrite and iron salts on methane formation by sewer biofilms has been studied recently at the University of Queensland, Australia. The used methodologies are presented here.

4.5.1 pH elevation

pH elevation through the addition of magnesium hydroxide ($Mg(OH)_2$) is commonly used as a means for reducing H_2S emission from wastewater to sewer air. When pH is elevated from a neutral level, which is typically in sewage, 8.5–9, the molecular hydrogen sulphide (H_2S) fraction of total dissolved sulphide is reduced, and as a result, it's transfer from the liquid to the gas phase is reduced.

Experiments were carried out on laboratory scale sewer systems consisting of both an experimental and a control reactor. Each reactor, with a volume of 1 l, was fed with domestic wastewater every six hours. During each pumping event, 1 l of wastewater was pumped into each reactor, resulting in a hydraulic retention time (HRT) of 6 hours. The 12-month experimental study was divided into four phases with conditions summarized in Table 4.4.

Table 4.4 Experimental conditions applied during the entire course of the pH study (Gutierrez et al., 2009)

Operational phases	Length (days)	Control Reactor	Experimental Reactor
1	0–40	No pH control 7.6±0.1	No pH control 7.6±0.1
2	51–110	No pH control 7.6±0.1	pH adjusted to 8.6±0.1 with 88 ml 0.05M NaOH in each pumping event
3	111–170	No pH control 7.6±0.1	pH adjusted to 9.0±0.1 with 120 ml 0.05M NaOH in each pumping event
4	171–332	No pH control 7.6±0.1	No pH control 7.6±0.1

4.5.2 Nitrite addition

Experiments were carried out both on laboratory systems and in practice. The methodology of both experiments is described underneath.

Laboratory experiments

The experimental set up for the experiments at laboratory scale (Jiang et al., 2010) is shown in Figure 4.3. This set up consisted of four air-tight reactors, namely R1 to R4, each with a volume of 0.75 l. The reactors were fed with sewage through a peristaltic pump every 6 hours, a typical sewage hydraulic retention time in sewers (Hvitved-Jacobsen, 2002). Every feed pumping event lasted for 2 minutes, delivering one reactor volume (0.75 l) of sewage into each reactor.

The experiments were conducted in three consecutive phases, namely the stabilization, dosing, and recovery phases. Reactors were operated without nitrite dosing to achieve similar sulphide and methane production activities during the stabilisation phase. During the dosing phase, R2-R4 received nitrite, while R1 was used as the control reactor (no nitrite dosage).

Reactors R2-R4 were injected with nitrite to reach concentrations of 40, 80, and 120 mg-N/l of wastewater, respectively. After the 24-day dosing phase, nitrite dosing to R2 - R4 was stopped and the reactors were allowed to recover for two and half months (Recovery phase).

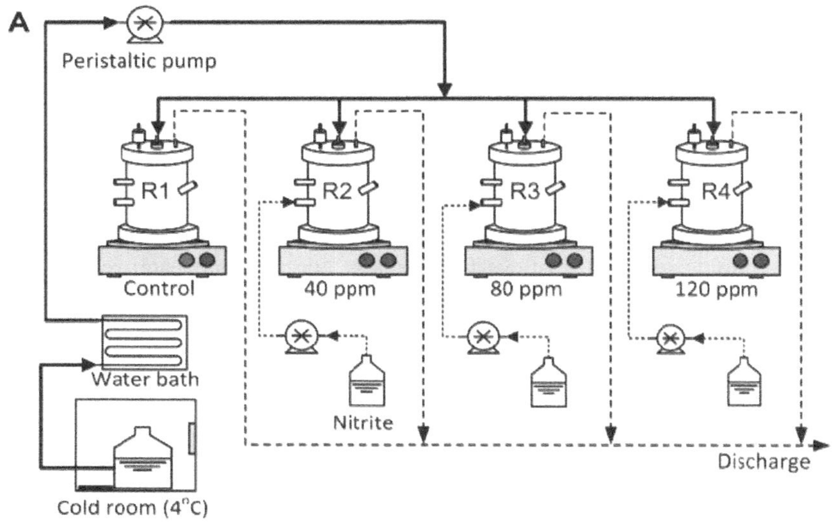

Figure 4.3 Schematic of the laboratory-scale rising main sewer reactors. R1 did not receive nitrite, and served as a control. R2, R3 and R4 received nitrite at 40, 80 and 120 mgN per l of wastewater fed, for a period of 24 days, which was followed by a 2.5 month recovery period.

Field trial

A field trial was conducted to test intermittent nitrite dosing as a means for sulphide and methane control in a real sewer line, UC09, located in Gold Coast, Australia. This sewer line has a length of 1080 m and a diameter of 150 mm. It receives primarily domestic wastewater with an average daily flow of approximately 200 m^3. The hydraulic retention time of sewage in the rising main varied between 1.7 and 5.7 hours during the period of the study.

Nitrite solution was dosed into the UC09 wet well manually during daytime only (8:00 AM to 7:00 PM) over three consecutive days. Before each pumping event, nitrite was added to the wet well, resulting in a concentration in sewage of 100 mg-N/l, a level determined based on the lab-scale reactor test results (Jiang et al., 2010). The dosage was stopped after three days. Therefore, the actual dosage time was 33 hours over a three day period.

Prior to nitrite dosing, three measurement campaigns were conducted to monitor methane production in the rising main over a period of seven days. Each campaign involved sampling wastewater hourly at both the pumping station wet well and 828 m downstream of the pumping station, for a period of 3–6 hours.

The samples were analyzed for dissolved methane concentrations using the method described in section 4.4.1. Similar measurement campaigns were also conducted 1, 4, 5, 10 and 13 weeks after nitrite dosing.

4.5.3 Iron salt addition

Two lab-scale rising main sewer systems fed with real sewage were operated for 8 months. One received Fe^{3+} dosage at 15 mg per l of wastewater (experimental system) and the other was used as a control (no Fe^{3+} dosage). (Zhang et al., 2009)

4.6 TOTAL CARBON FOOTPRINT WWTP

In the research performed in the Netherlands the emission of N_2O and CH_4 was measured at the same three WWTPs. Based on these measurements it was possible to determine the contribution of N_2O and CH_4 to the total carbon footprint of a WWTP including the use of electricity and natural gas. To determine the total carbon footprint of a WWTP all sources were converted to CO_2 equivalents. The conversions factors that have been used in this case are summarised in Table 4.5.

Table 4.5 Conversion factors for climate footprint calculations used in the Netherlands.

	Conversion factor	Unit
Electricity	0.67	kg CO_2/kWh
Natural gas	1.8	kg CO_2/Nm^3
N_2O	298[1)]	kg CO_2/kg N_2O
CH_4	25[1]	kg CO_2/kg CH_4

[1]GWP AR4: Adjusted GWP in IPCC Fourth assessment report, 2007 (IPCC, 2007).

The conversion factors mentioned for electricity and natural gas in Table 4.5 are specific for the situation in the Netherlands. In any other country, other factors apply depending on the energy mix used.

The total carbon footprint analysis did not account for the CO_2 required to produce chemicals nor did it account for the emission of greenhouses gases emitted at the sites were the sludge was further handled (i.e. incinerated or composted).

Chapter 5
Results

5.1 N_2O EMISSION FROM WWTPS

5.1.1 Emission of N_2O

The emission of N_2O as measured at all the different WWTPs in Australia, France, USA and the Netherlands is summarised in Table 5.1.

Table 5.1 Overview N_2O emission at the surveyed WWTPs around the world (values presented in bold are reported outliers; for more details see technical report (GWRC,2011).

Country	WWTP	Sample round	Emission (kg N_2O/kg $TKN_{influent}$)	Emission (kg N_2O-N/kg $TKN_{influent}$)
Australia	1 – Ox. Ditch	1.	0.005	0.003
		2.	0.003	0.002
		3.	0.006	0.004
		4.	0.003	0.002
	2 – Johannesburg	1.	0.016	0.010
		2.	0.027	0.017
		3.	0.011	0.007
	3 – SBR	1.	0.009	0.006
		2.	0.011	0.007
		3.	**0.050**	**0.032**
	4 – MLE (1)	1.	0.019	0.012

(Continued)

Table 5.1 Overview N_2O emission at the surveyed WWTPs around the world (values presented in bold are reported outliers; for more details see technical report (GWRC,2011)(*Continued*).

Country	WWTP	Sample round	Emission (kg N_2O/kg $TKN_{influent}$)	Emission (kg N_2O-N/kg $TKN_{influent}$)
	5 – MLE (2)	1.	0.014	0.009
		2.	**0.047**	**0.030**
		3.	0.006	0.004
		4.	**0.055**	**0.035**
	6 – MLE (3)	1.	**0.176**	**0.112**
		2.	0.008	0.005
		3.	0.005	0.003
	7 – A^2O	1.	0.011	0.007
		3.	0.006	0.004
France	WWTP 1		<0.0002	<0.0001
	WWTP 2		<0.0002	<0.0001
	WWTP 3		0.003	0.0017
	WWTP 4		0.002	0.0011
USA	Separate-stage BNR	1. (15°C)	0.0005	0.0003
		2. (23°C)	0.0002	0.0001
	Four-stage Bardenpho	1. (14°C)	0.0025	0.0016
		2. (23°C)	0.0094	0.006
	Step-feed BNR 1	1. (19°C)	0.025	0.016
		2. (25°C)	0.0097	0.0062
	Step-feed non-BNR	1. (17°C)	0.0028	0.0018
		2. (26°C)	0.028	0.018
	Separate centrate[1]	1. (30°C)	0.0038	0.0024
		2. (34°C)	0.0085	0.0054
	Plug-flow 1	1. (11°C)	0.0063	0.004
		2. (23°C)	0.0064	0.0041
	Plug-flow 2	1. (11°C)	0.0097	0.0062
		2. (22°C)	0.0014	0.0009
	MLE 1	1. (26°C)	0.0011	0.0007
	MLE 2	1. (26°C)	0.0009	0.0006

(*Continued*)

Table 5.1 Overview N_2O emission at the surveyed WWTPs around the world (values presented in bold are reported outliers; for more details see technical report (GWRC,2011)(*Continued*).

Country	WWTP	Sample round	Emission (kg N_2O/kg $TKN_{influent}$)	Emission (kg N_2O-N/kg $TKN_{influent}$)
	Step-feed BNR 2	1. (29°C)	0.024	0.015
	Oxidation ditch	1. (19°C)	0.0005	0.0003
	Step-feed BNR 3	1. (24°C)	0.0008	0.0005
Netherlands	Papendrecht	1. (19°C)	0.00063	0.00040
	Kortenoord	1. (19°C)	0.00075	0.00048
	Kralingseveer	1. (18°C)	0.0066	0.0042
		2. (10°C)	**0.096**	**0.061**

[1] Not a typical WWTP

Based on the emission data presented in Table 5.1 the following observations can be made:

- The emission of N_2O varies greatly among WWTPs measured, the lowest emission was smaller than 0.0001 kg N_2O-N/kg $TKN_{influent}$ the highest emission was as high as 0.112 kg N_2O-N/kg $TKN_{influent}$.
- The emission at the different WWTPs per country differs greatly depending on configuration and operations.
- The emission varies in between seasons as shown by the results from the USA and the Netherlands

Besides the observed variations in N_2O emission mentioned above, a strong variation was observed during a day at one WWTP. An example of this from the WWTP Kralingseveer, the Netherlands is presented in Figure 5.1. It can be observed that the emission of N_2O varied with the influent flow. This suggests that the emission of N_2O is related to the variation in sludge load. A similar relation was found in the USA study, where the diurnal variability observed could be linked to diurnal variations in influent N-loading (Ahn et al., 2009). An example of this is presented in Figure 5.2.

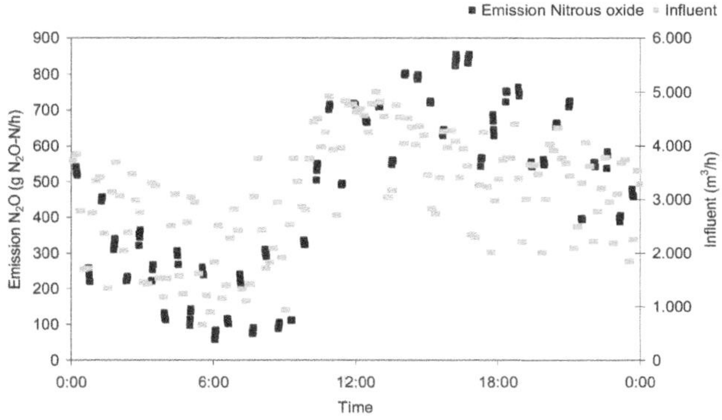

Figure 5.1 N_2O Emission and influent flow on 18-10-2008 at WWTP Kralingseveer.

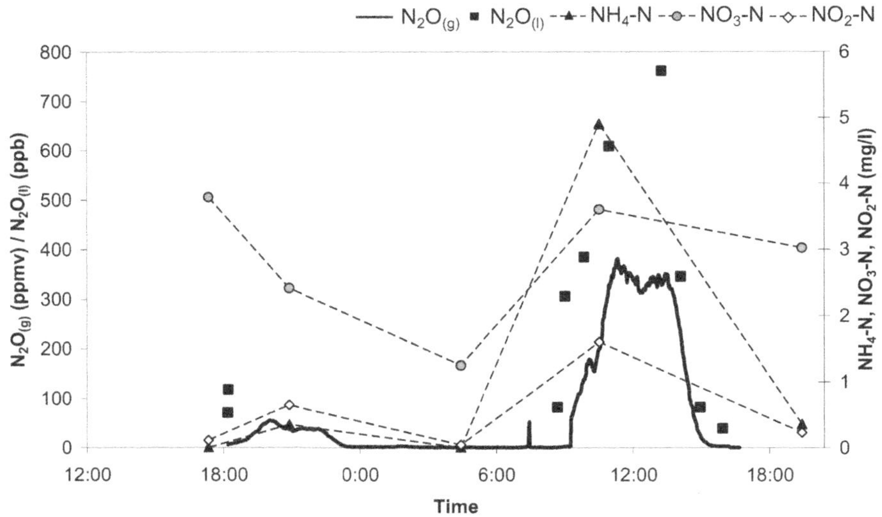

Figure 5.2 Diurnal variability in gaseous N_2O concentrations measured from an aerobic zone of the full-scale step-feed BNR process.

5.1.2 Origin of N_2O emission

N_2O can be formed and emitted from anoxic and aerobic zones, i.e. N_2O can be formed during both denitrification and nitrification. Based on the measurements at the 12 WWTPs in the USA the emission of N_2O was in general higher from aerated zones than from non-aerated zones as can be seen from Figure 5.3 (note y-axis is logarithmic).

This indicates that the nitrification could be responsible for the formation of N_2O. However, based on these data the distinction between stripping of N_2O (from anoxic zones) or formation in the aerobic zone cannot be made. More information on this point can be obtained when measurements on liquid and gas phase N_2O are available per reactor zone. These measurements have been performed at different WWTPs in the USA. The results from one of them are presented in Table 5.2.

In Table 5.2 there is a strong increase in N_2O concentration at the transition from the anoxic zone to the first aerobic zone. At this point it is thought that the nitrification was the main process that contributed to the N_2O formation, however the contribution of the denitrification cannot completely be excluded here. At the transition from the first aerobic zone to the second zone another strong increase in N_2O concentration can be observed, which is a strong indication that nitrification was responsible for the formation of N_2O. This observation is supported by a measurement in a complete stirred reactor (WWTP1; France) where the N_2O concentration in the gas phase was monitored before and after the start of the aeration. The results of these measurements are presented in Figure 5.7.

From Figure 5.4 it can be observed that a first peak of N_2O occurred at the start of the aeration, which is most likely formed earlier during denitrification. A second larger peak occurred during the aeration period. This indicates that the emitted N_2O is formed by nitrifying bacteria.

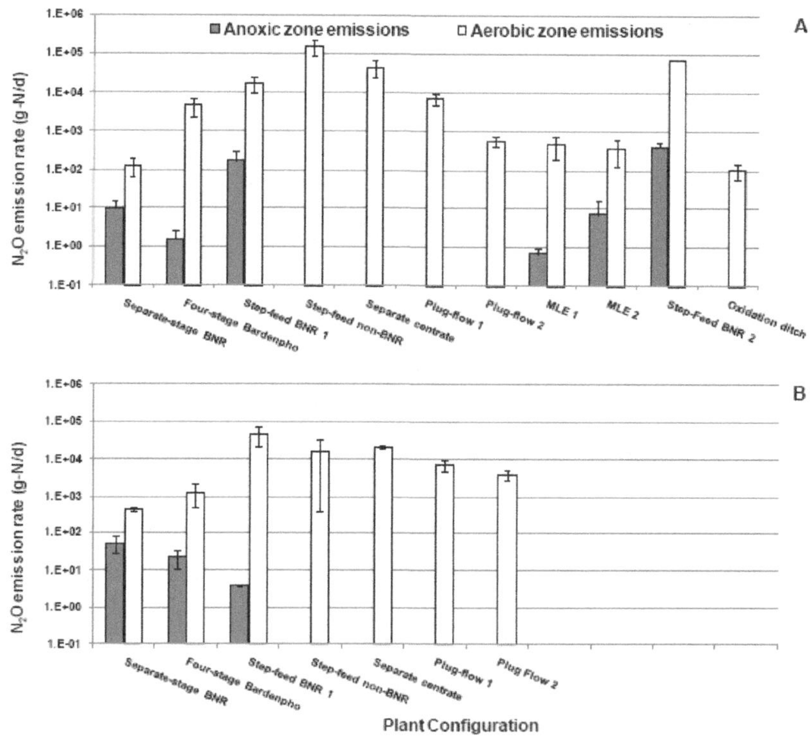

Figure 5.3 N_2O emissions from aerobic and anoxic zones in different WWTPs measured at high (A) and low (B) temperatures. Specific temperatures described in Table 5.1. Step-feed BNR 3 is not included since the emissions from the covered aerobic and anoxic zones could not be distinctly measured.

Table 5.2 Spatial profile of gaseous N_2O concentrations and typical activated sludge variables in a Step-feed BNR process showing individual sampling locations. Results are from discrete sampling over a period of 30 minutes at each sampling point. Arrows indicate wastewater flow. Shaded and unshaded boxes represent non-aerated and aerated zones, respectively. Gaseous N_2O concentrations are expressed as avg. ± sd. of 30 measurements.

Species	Anoxic →	Aerobic 1 →	Aerobic 2 →
NH_4^+ (mg-N/l)	14	12 ± 5	1.5 ± 0.71
NO_2^- (mg-N/l)	0.00 ± 0.00	0.00 ± 0.00	0.00 ± 0.00
NO_3^- (mg-N/l)	0.85 ± 0.10	2.7 ± 0.35	10 ± 0.21
DO (mg O_2/l)	0.10	2.3	4.2
Aqueous N_2O (μg/l)	55	190	570
Gaseous N_2O (ppmv)	1.5 ± 0.14	16 ± 0.27	23 ± 0.67

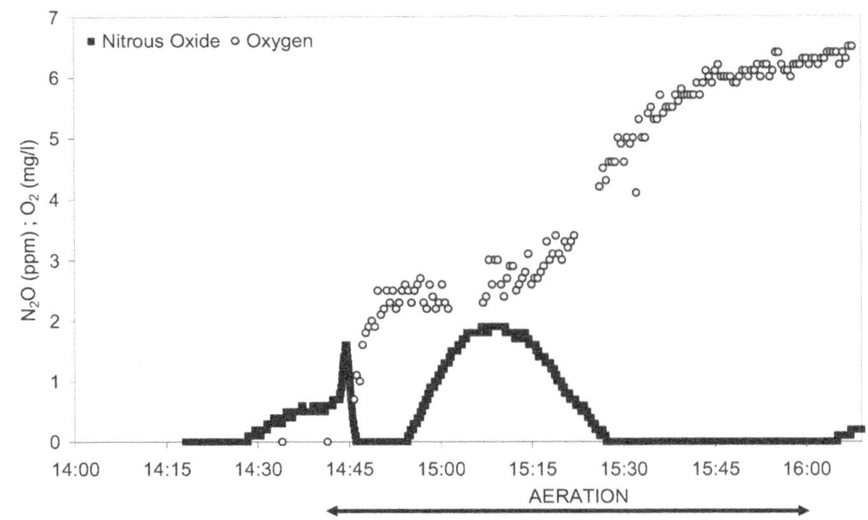

Figure 5.4 N_2O emission during nitrification in WWTP1 (France); date 06/2007.

5.1.3 Process parameters of influence

Knowledge on the process parameters that influence the formation and emission of N_2O can help to define measures to control the formation and emission of N_2O. Based on the results in the USA, multivariate regression modelling[1] was performed on all the collected data. This modelling was performed for both the aerobic and the anoxic zones.

The factors that positively correlated with N_2O emissions from aerobic zones were NH_4^+, NO_2^- and DO concentrations (isolated effect), and NH_4^+ and NO_2^- concentrations (combined effect).

The factors that positively correlated with N_2O emissions from anoxic zones was the DO and NO_2-N concentration (combined effect).

The positive correlation between N_2O emission and the concentration of nitrite in the liquid phase, as found in the USA, is supported by the results found in Australia, as can be observed from Figure 5.5A. Although the number of data points is limited, there appears to be a threshold value at approximately 0.3–0.5 mg NO_2-N/l at which the generation factor increases sharply.

Figure 5.5 shows also the emission of N_2O as function of the effluent total nitrogen (B) and a-recycle rate (C). From these two last figures it can be observed that two WWTPs (Oxidation ditch and Johannesburg) have very high a-recycle rates[2] and correspondingly low effluent TN concentrations. These two WWTPs are also among those with the lowest N_2O emission, which indicates that the level of recirculation is a process parameter that influences the level of N_2O emission.

Although a large variation in N_2O emission was observed at the different temperatures (see Table 5.1) no correlation could be found between N_2O emission and temperature (based on multivariate regression modelling results USA).

[1]Details on the used methodology and results can be found in Chandran, 2010.

[2]Recycle between aerobic and anoxic zone, as a multiple of the average influent flowrate.

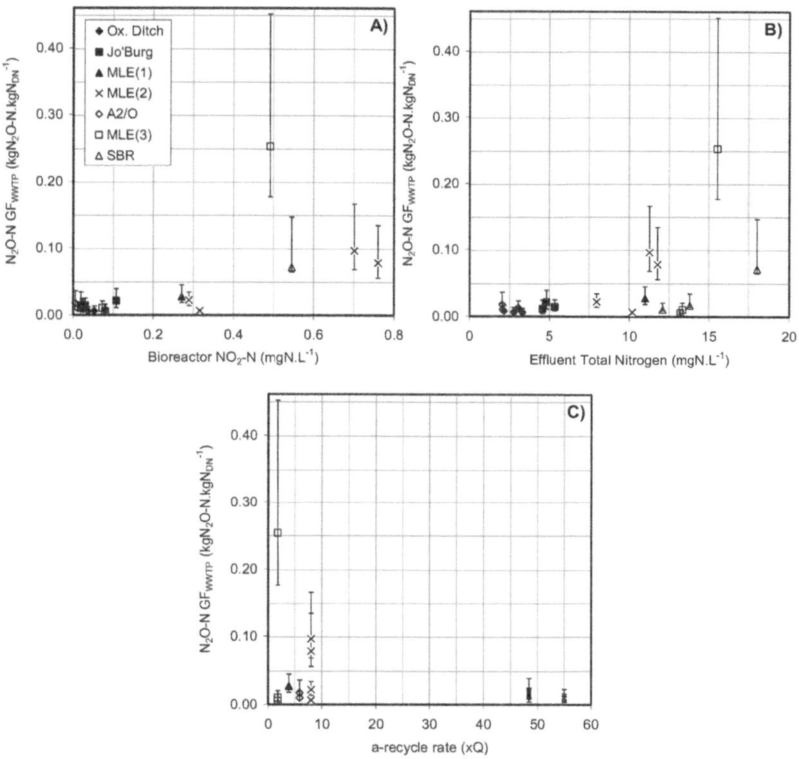

Figure 5.5 Net N_2O-N generation factor, GF_{WWTP}, in each sampling round, plotted against (A) bulk bioreactor nitrite-N concentration, (B) effluent total nitrogen, and (C) a-recycle rate (aerobic → anoxic) as a multiple of the average influent flowrate, Q.

5.2 CH_4 EMISSION FROM SEWERS

5.2.1 Liquid phase

The results of one of the measuring campaigns at the CO16 raising main in Australia are presented in Figure 5.6.

From Figure 5.6 it can be observed that already some methane is present in the pumping station (1–2 mg/l), and that the concentration increased further downstream, reaching approximately 9.0 mg/l. Furthermore it can be observed that the methane concentration varied considerably with time at the 500m and 1100m locations, likely due to the variation of HRT of the wastewater samples collected. By taking all results together, the average methane production rate was calculated as 0.7 $mg \cdot l^{-1}\ hr^{-1}$ or in terms of sewer biofilm area 1.3 $g \cdot m^{-2} \cdot d^{-1}$.

In another measuring campaign, the 1100m location was sampled half-hourly during 4:15 a.m. and 10:00 a.m. The results are presented in Figure 5.7.

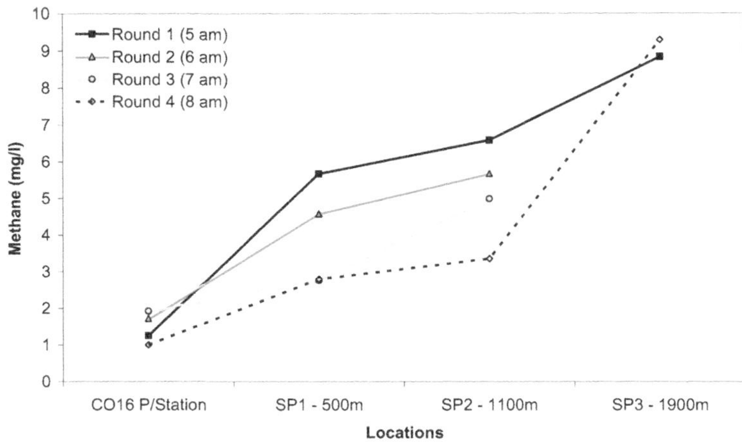

Figure 5.6 Methane data measured at four locations along CO16. The average wastewater temperature was 23.5°C (Foley et al., 2009).

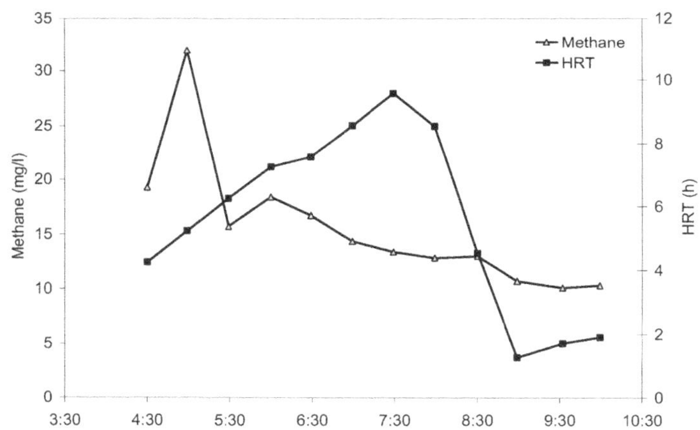

Figure 5.7 Methane data measured at CO16 at 1,100m in another campaign. The average wastewater temperature was 22.5 °C.

The methane concentrations measured in this measuring campaign were approximately three times higher than those presented in Figure 5.6. This difference could not be fully explained by the differences in HRT (4.0–9.5 hr in this case in comparison with 1.5–7.3 hr in first case). The average methane production rate was calculated as 3.1 $mg \cdot l^{-1}.hr^{-1}$ or 5.5 $g \cdot m^{-2} \cdot d^{-1}$, three times higher than the values obtained in the previous case. The average wastewater temperature was 22.5°C, which was similar to that in the previous case. The reason for the much higher methane production in this case is not known. This could be related to the COD (Chemical Oxygen Demand) concentration, which was unfortunately not measured during either measurement campaign.

5.2.2 Gas phase

The results of the measuring campaigns at 64 (presented 59) lift stations in the USA are presented in Figure 5.8.

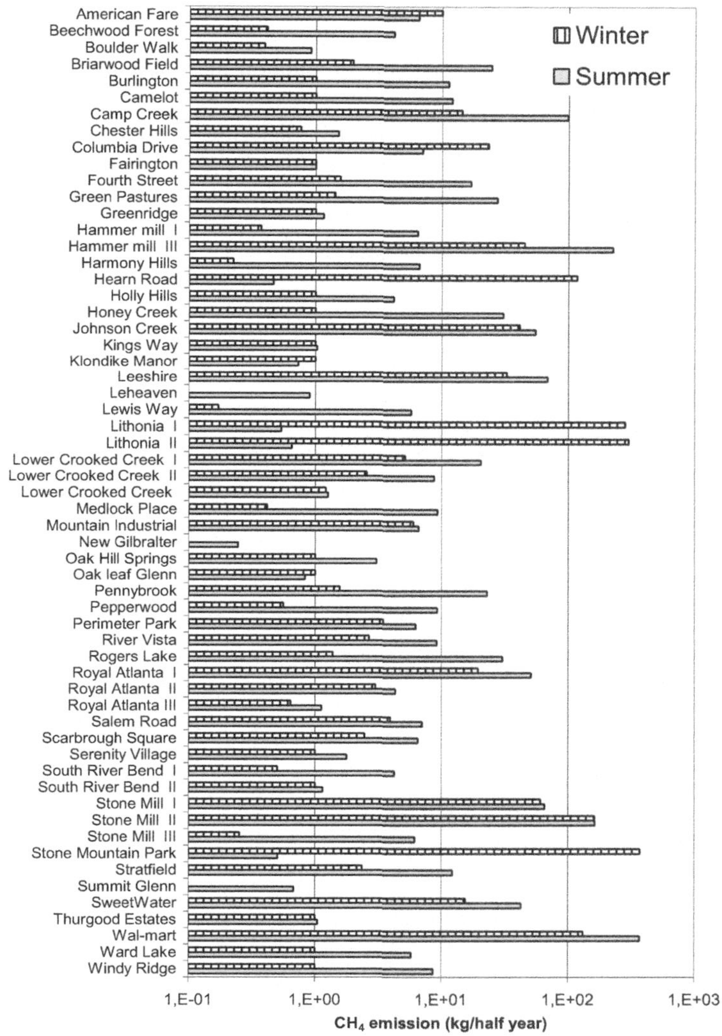

Figure 5.8 Instantaneous CH_4 readings during winter and summer monitoring at 59 lift stations in DeKalb County, Georgia, USA (note x-axis is logarithmic).

Figure 5.8 shows the calculated semi-annual mass emissions calculated from the winter and summer grab samples and the volumetric displacement of each station during each semi-annual period. Most of the summer emissions are higher than the winter emissions; likely because more methane is formed and methane solubility is less at higher temperatures.

5.2.3 Mitigation strategies

The results of three mitigation strategies will be presented in this section, which are pH elevation, nitrite and iron salt addition.

pH elevation

The effect of pH elevation on the VFA[3] and methane production in a laboratory scale sewer reactor is presented in Figure 5.9.

From Figure 5.9 it can be observed that methanogenic activities developed in the control reactor within three months after the reactor start-up, while no significant methanogenic activities were detected in the experimental reactor until normal pH was resumed. The results suggest that elevated pH at 8.6–9.0 suppressed the growth of methanogens.

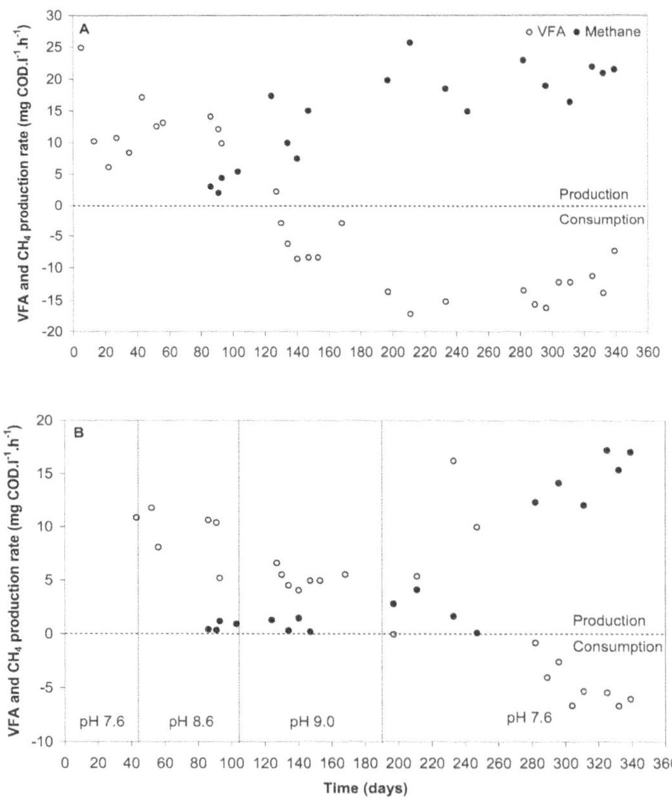

Figure 5.9 Evolution of the VFA (○) and methane production rates (●) in the control (A) and experimental (B) reactors. Note that methane production rates have a unit of $mgCOD \cdot l^{-1} \cdot hr^{-1}$ in this figure, rather than $mgCH_4 \cdot l^{-1} \cdot hr^{-1}$, for direct comparison with the VFA production/consumption rates. 1 $mgCH_4$ = 4 mg COD.

---

[3]Volatile fatty acids.

Elevated pH conditions also reduced the activity of fermentative bacteria (FB) in the reactors. Prior to the development of the methanogenic activity, a stable VFA production rate of 11.5 ± 1.2 mgVFA COD·l^{-1}·hr^{-1} was observed in the control and experimental reactors. The FB activity decreased to 5.1 ± 0.4 mgVFA-COD·l^{-1}·hr^{-1} in the experimental reactor under pH 9.0 (with negligible methane formation). This indicates that the fermentation activity was reduced by 54% at pH 9.0 in comparison to pH 7.6.

Furthermore, it was observed that the activity of sulphate reducing bacteria was reduced by 30–50% respectively at pH 8.6 and pH 9.0. The results further showed that pH elevation not only reduced the H_2S transfer but also its production by sewer biofilms. When normal pH was resumed, it took approximately 2 months for the SRB (sulphate reducing bacteria) activity to fully recover (Gutierrez et al., 2009).

Nitrite addition

The impact of nitrite addition on the methanogenic and sulphate reducing activities were studied on both laboratory-, and full scale. The results of the laboratory experiments are presented in Figure 5.10.

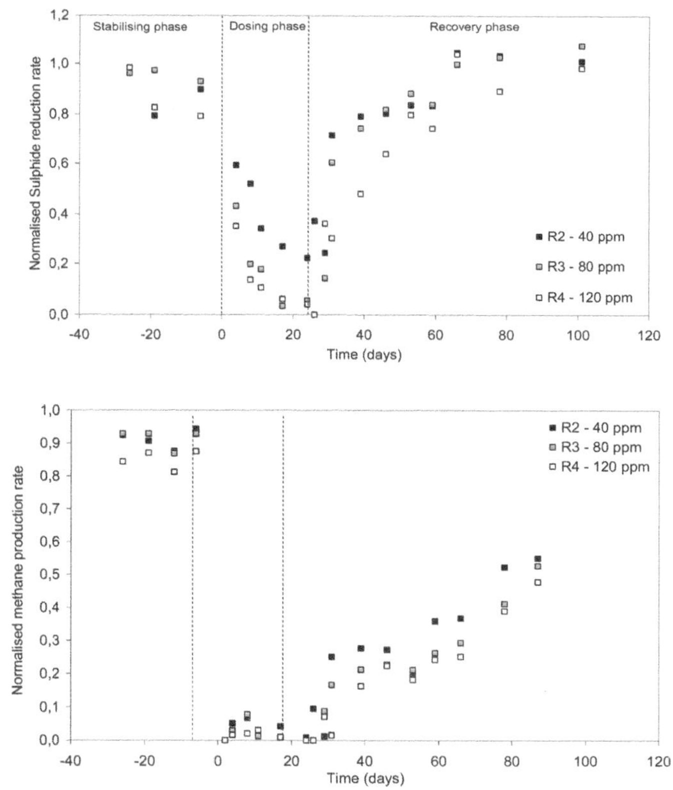

Figure 5.10 Normalized methane production rate in nitrite-dosed reactors: R2 (○), R3 (□), and R4 (Δ), relative to the control reactor R1 (i.e. the activity measured for R1 on each measurement day was considered 100%, to eliminate the rate variation caused by changes in wastewater quality). Nitrite addition was initiated on Day 0 and finished on Day 24.

From Figure 5.10 it can be observed that:

- The methane [4] production in dosed reactors was reduced to negligible levels within two days after the initiation of nitrite dosage.
- The different levels of nitrite addition did not cause any difference in inhibition.
- Complete methanogenic inhibition was achieved with the lowest nitrite concentration used in the experiment, i.e. 40 mg-N/l
- The recovery of methane production proceeded at very similar rates in all cases.
- The recovery process was almost linear during the whole recovery phase.
- Only less than 60% of recovery was achieved after two-month of recovery.
- Suppression of the sulphide production was also achieved with nitrite dosage at 80 mg-N/l and 120 mg-N/l.
- The recovery of the sulphide production was faster than the recovery of the methane production.

The results of the field trial at the real sewer line UC09 are presented in Figure 5.11, which shows the concentrations of methane in the wet well (0m) and at the point 828m downstream.

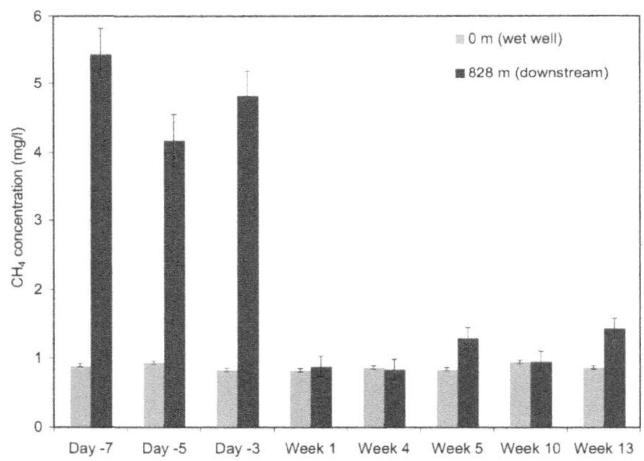

Figure 5.11 Daily average methane concentration at 828m and at the pumping station wet well. Day -7, -5, and -3 were during the baseline period. Nitrite was added intermittently at 100 mgN/L during Day 0–2 for a total dosing time of 33 hours over a 3-day period. Week 1 to 13 imply 1–13 weeks after the dosage. Methane concentrations were based on manual sampling and offline measurements (3–6 samples were taken on each sampling day). The error bars shown are standard errors (n = 3–6).

Complete suppression of methane production by nitrite addition was observed, as indicated by the nearly identical methane concentrations in the pumping station and at 828 m. One month after terminating nitrite dosage, methane concentration at 828 m remained at a level similar to that measured in the wet well, indicating that the sewer biofilm ceased to produce methane in this period. The activity increased only marginally in the following two months.

[4] The methanogenic activity was measured in the absence of nitrite through batch tests using fresh sewage without nitrite addition in all cases.

In general, the field trial confirmed the results of the laboratory study, that nitrite has a long-term toxic effect on methanogens (and sulphate reducing bacteria – data not shown) in anaerobic sewer biofilms. Both the field and laboratory results collectively suggest that nitrite could be applied intermittently to achieve sulphide and methane control in sewers.

Iron salt addition

In the experiments where iron salts were added to laboratory scale rising main sewer systems, Fe^{3+} dosage was found to significantly inhibit (next to precipitation with sulphide) sulphate reduction and methane production by sewer biofilms. The sulphate reduction rate was found to be inhibited by 40–60%, while the methanogenic activity was inhibited by 50–80%. The rate data were supported by significantly lower methane concentrations at the end of the experimental rising main system (Zhang et al., 2009). The mechanisms responsible for the inhibition observed are yet to be fully understood.

5.3 CH_4 EMISSION FROM WWTPS

5.3.1 Emission of CH_4

The emission of CH_4 as reported at the four WWTPs in France and the three WWTPs in the Netherlands are presented in Table 5.3.

Table 5.3 Emission of CH_4 from WWTPs in France and the Netherlands

Country	WWTP	Emission ($kgCH_4$//$kgCOD_{influent}$)	Emission (kg CH_4-COD/kg $COD_{influent}$)
France	WWTP 1	< 0.0001	< 0.0004
	WWTP 2	< 0.0001	< 0.0004
	WWTP 3	0.0004	0.0016
	WWTP 4	0.0003	0.0012
The Netherlands	Papendrecht	0.0087	0.035
	Kortenoord	0.0053	0.021
	Kralingseveer October	0.012	0.048
	Kralingseveer February	0.008	0.032

Based on Table 5.3 the following observations can be made:

- The highest emission of methane occurred at WWTP Kralingseveer in October. This could be related to the presence of a sludge digester at this WWTP where at Papendrecht and Kortenoord a sludge digester is absent.

- In February the emission at Kralingseveer was lower than in October. This could be related to the temperature of the water which was 19°C in October and around 10°C in February. At lower temperatures less methane will (probably) be produced and more methane will be dissolved. For this reason methane could have left the WWTP via sludge or the effluent. However, the concentration of soluble methane in both streams was not determined.
- The emission factors for CH_4 differed (in some case higher; in other cases lower) than the emission factor[5] currently used in the Netherlands to estimate the emission of CH_4 from WWTPs.

5.3.2 Origin of emission

At the three WWTPs in the Netherlands, the total CH_4 emission was determined based on grab samples from all present process parts. In this way the origin of the methane emission could be determined. The results for the two WWTPs without sludge digestion are presented in Figure 5.12.

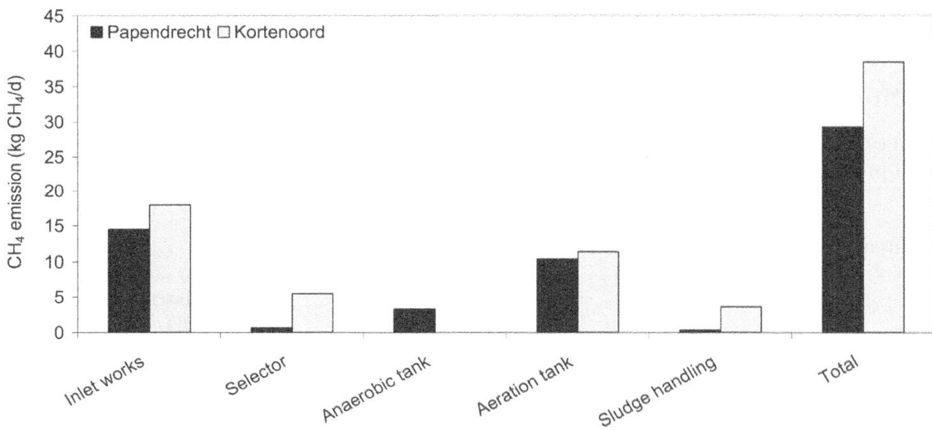

Figure 5.12 Origin of methane emission at the WWTPs in Papendrecht and Kortenoord.

The emission of CH_4 at the two WWTPs primarily occurred from the inlet works and the aeration tanks. The contribution of the sludge handling sites was minimal. In both cases, the contribution from the inlet works formed almost half of the total CH_4 emission. The contribution from the aeration tank was 36% in Papendrecht and 30% in Kortenoord. It should be noted that some of the CH_4 emitted at a WWTP could have originated from sewers. CH_4 formed and subsequently transported to the treatment plants can be stripped at the inlet works and the subsequent process units.

For both measurement periods at Kralingseveer the origin of the CH_4 emission is presented in Figure 5.13.

At WWTP Kralingseveer the ventilated air from all anaerobic parts is treated separately in a compost filter. After this treatment, the air is reused in the aeration tanks 1 and 2. For this reason the total CH_4 emission after the anaerobic parts is presented as well in Figure 5.13.

---

[5] For the national inventory reports 0.007 kg CH_4/kg $COD_{influent}$ are used for WWTPs without sludge digestion and 0.0085 kg CH_4/kg $COD_{influent}$ for WWTPs with sludge digestion (VROM, 2008).

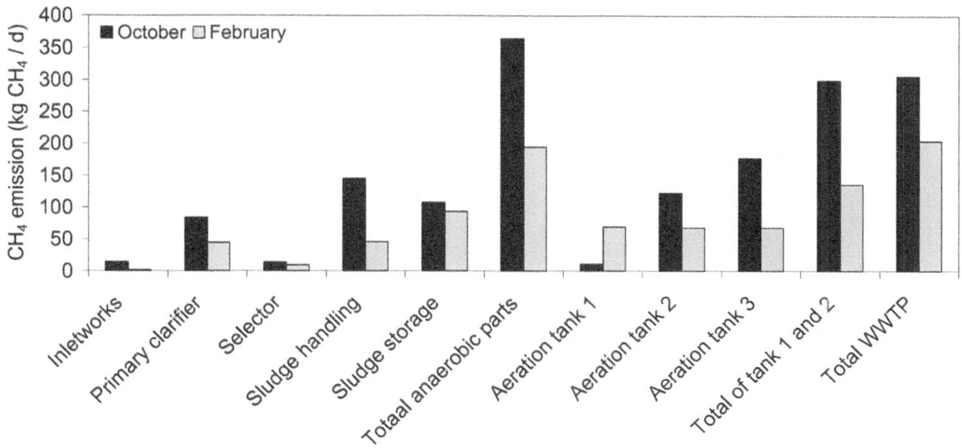

Figure 5.13 Overview of the CH_4 emission at WWTP Kralinseveer during the measurement campaign in October (2008) and February (2009).

From Figure 5.13 it can be observed that the emission of CH_4 from all anaerobic parts is larger than the total CH_4 emission from aeration tanks 1 and 2. This implies that some CH_4 is lost or converted during treatment in the compost filter or in the old aeration tanks. The loss of CH_4 in October was 92 kg CH_4/d and for February the loss was 59 kg CH_4/d.

The main parts that contributed to the emission of CH_4 were the primary clarifier, the sludge handling and storage units. The contribution of the primary clarifier was 23% in both periods. For the sludge handling the contribution was 45% in October and 23% in February. Finally the sludge storage contributed 28% in October and 48% in February.

5.4 TOTAL CARBON FOOTPRINT

The contribution of each greenhouse gas to the total carbon footprint of the three WWTPs from the Netherlands is presented in Figure 5.14.

From Figure 5.14 it can be observed that:

- Electricity forms the major contributor to the total carbon footprint at WWTPs with a low N-sludge load (Papendrecht and Kortenoord).
- At the WWTP (Kralingseveer) with a higher N-sludge load the contribution of N_2O can be substantial.
- The contribution of CH_4 to the total carbon footprint depends on the presence of a sludge digester. In the last case (Kralingseveer) the contribution of CH_4 can be substantial.

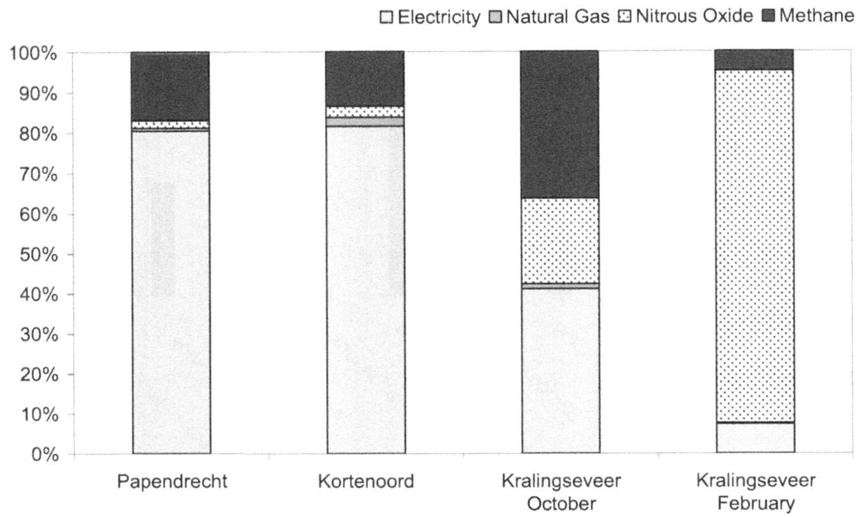

Figure 5.14 Total carbon footprint of WWTP Kortenoord (A; the Netherlands) and WWTP Kralingseveer October (B; the Netherlands).

Chapter 6
Discussion

6.1 METHODOLOGY

6.1.1 N_2O emission

The research presented in this report used different methodologies to determine the emission of N_2O from WWTPs. The weak and strong points of these monitoring technologies are summarised in Table 6.1.

Table 6.1 Strong and weak points methodologies to determine N_2O emission from WWTPs regarding variability and origin (+ refers to be suitable, +/- refers to be less suitable, −refers to be not suitable).

Methodology	Variability	Origin/process parameters of influence
Liquid phase measurements/mass balance (Australia)	-	+
Gas hood (France/USA)	+/-	+
Gas phase covered WWTPs (The Netherlands)	+	+/-

To determine the N_2O emission from uncovered WWTPs different methodologies were developed. In Australia a method was developed based on liquid phase measurements for individual reactor zones. Based on mass transfer coefficients a complete mass balance over the different zones and the complete WWTP could be made. In this way the emission of N_2O was determined, but it was also possible to differentiate between N_2O emission and generation. However, this method does not allow for sufficient insight in the variation of N_2O emission as the method is based on grab samples.

The methodologies developed in the USA and France were based on gas hoods measurements. Those gas hoods were placed at the different zones of a WWTP and in case of the USA, the gas hoods measurements were combined with liquid phase measurements. By placing the gas hoods at the different zones of a WWTP and combine those with the liquid phase measurement it was also in the USA possible to differentiate between N_2O emission and generation. On the other hand, the gas-hood measurements are limited by the dimensions of the hood itself. Additionally, heterogeneities in the emission gas flow rate could also contribute to the variability in the actual emissions, which needs to be considered. To calculate the total emissions load from a plant, both the N_2O concentrations and advective gas flow must linearly be scaled up from the hoods to the entire activated sludge bioreactor. Alternately, multiple measurements need to be done at different locations in the activated sludge reactor to capture the spatial variability in emissions.

The methodology used in the Netherlands was based on the analyses of the total off gas of covered WWTPs. The advantage of this method is that it captures the total emissions given the variability of N_2O emission in time and space, but with this methodology it is more difficult to differentiate between N_2O emission and generation.

To finally estimate and control the emission of N_2O from WWTPs, the use of all available methods will be required.

6.1.2 CH_4 emission

In the research performed in Australia, no CH_4 emission data were calculated. In the USA, the emission of methane was calculated from unventilated lift stations. For this gas flow needed to be determined. This was done via the active volume of each lift station in combination with the pumping rate of each of the pumps. It is hereby assumed that the amount of air breathed in and exhausted out during each pumping cycle is approximately equal to the active volume inside the wet well or the volume of raw sewage pumped during each cycle. However, the methodology used to determine the gas flow is subject to a number of limitations, namely:

- Dispersion of CH_4, which is lighter than air, is not accounted for. Many stations cycle only a few times per day and those stations likely lose a considerable portion of the evolved CH_4 to the atmosphere in between cycles. This phenomenon would result in under-reporting of actual emissions.
- Dispersion effects can be very significant at some of the larger pumping stations, like LCC-3. LCC-3 has a large, uncovered forebay with significant interchanges of outside air. On calm days the forbay air changes could be 1 to 5 times per hour and could increase to as much as 10 or 30 times per hour on windy days. This limitation, combined with the fact that these stations have the highest volume and mass throughput which should produce more CH_4, could represent a very significant source for under-reporting.
- During a pumping cycle, a significant volume of outside air is drawn into the wet well. If a sampling event occurred soon after such a dilution, the applied concentrations could under-report the actual CH_4 emissions.

These limitations show that there is a need for a good strategy to measure loads of methane emitted to the air from sewerage systems.

6.2 N_2O EMISSION

6.2.1 Emission

The results on the emission of N_2O from WWTPs showed that there exists a great variability among different WWTPs and at the same WWTP during different seasons or throughout the day. This great variability results from the fact that the formation of N_2O is very complex and can be executed by both autotrophic and heterotrophic bacteria. Those bacteria show a dynamic response to changing influent loads and process conditions resulting in a variable formation of N_2O in time and place. Because of the observed variability in N_2O emission the use of a generic emission factor to estimate the emission from a specific WWTP is inadequate. To determine the emission of N_2O from a specific WWTP measurements are required.

6.2.2 Origin

N_2O can originate from denitrification and nitrification processes. Although the general assumption was that denitrification is likely the major source, the measurements and also several associated laboratory studies (e.g. Burgess et al., 2002) indicate that likely nitrification is a more important source of N_2O in the wastewater treatment plants.

This is observed from measured higher liquid N_2O levels in aerated zones compared to anoxic zones (see Table 5.2). Also when anoxic sludge gets aerated observations generally showed an initial sharp peak of N_2O emission likely due to stripping, whereas later a much higher amount of N_2O is emitted likely due to N_2O formation and stripping (see Figure 5.4).

6.2.3 Process parameters influence

Nitrite was measured in the USA plants; there it showed a strong correlation with N_2O emission. The results from Australia supported these findings as could be observed from Figure 5.5A.

The stimulating effect of nitrite (especially at low concentrations 0–2 mgN/l) has also been shown by several laboratory investigations (e.g. Tallec, 2006). Therefore it is likely that all factors (lower DO, sudden changes in ammonium load) that lead to nitrite build up in a treatment plant will also enhance N_2O emissions.

Ammonium is a substrate for N_2O formation by nitrifying bacteria. The correlation found in the US study between ammonium levels and N_2O emission is in accordance with this fact and the theory that N_2O is mainly derived from nitrification. The observations of correlations between daily variations in ammonium load and N_2O emission are in line with ammonium substrate utilization but are also in line with laboratory results showing that when nitrifying bacteria increase their nitrification rates they produce N_2O (Yu et al., 2010).

For other factors (e.g. COD availability) reported in literature to have a potential impact on N_2O emissions were not supported by direct observations that they are important. Given the level of knowledge, however, it should not be concluded that such factors do not play a role.

6.2.4 Implications of gained knowledge

Based on the above, it becomes clear that the formation of N_2O increases in aerobic zones at increasing levels of NH_4^+, NO_2^- and DO (but also at low levels). This implies that if high concentrations of these constituents can be avoided in practice, the risk of N_2O emission will be reduced. These high concentrations can be avoided in:

- systems that approach "ideal" well-mixed conditions (i.e. high recycle rates). In this way concentrations of intermediates from nitrification-denitrification including NO_2-N and NO are diluted, thereby reducing their inhibitory effect (Casey et al., 1999a,b). This is confirmed by the general results from all studies that WWTPs with the highest recycle rates and with the lowest TN effluent concentrations belonged to the WWTPs with the lowest N_2O emission.

- systems that avoid over – aeration i.e. have a rapidly responding DO control systems. In this way high levels of DO in aerobic and anoxic zones can be avoided
- systems that are equipped with flow equalization. Such systems minimize the peaking factor of influent nitrogen (ammonium) loading to the activated sludge system.

With the presented insight, it has been possible to estimate the risk of N_2O emissions from a specific WWTP based on the risk matrix presented in Table 6.2.

Table 6.2 Risk matrix to determine risk level of N_2O emission.

Risk on N_2O Parameter	High	Medium	Low
Effluent total nitrogen (mg/l)	>10	5–10	<5
Range in N-concentration in plant	H	M	L
Load variations	H	M	L
Maximum nitrite concentrations anywhere in plant	>0.5*	0.2–0.5	0.2

*Risk does not increase at nitrite concentrations higher than 2 mg N2O-N/l.

In summary, it can be concluded that in systems with extensive nitrogen removal the formation of N_2O is highly minimized. In other words, there is no conflict between water quality and air quality; rather they go hand in hand. This implies that systems that are not designed for (complete) nitrogen removal exhibit a high risk of N_2O emission. At temperatures of 20°C, an aerobic SRT of 2–3 days is already sufficient to get nitrification. In these high-loaded systems, nitrite easily accumulates resulting in greater N_2O formation.

6.2.5 Future research

The results of the presented research from Australia, France, the USA and the Netherlands already clearly identified key process parameters that influence the emission of N_2O from WWTPs. This knowledge presents a good starting point for defining mitigation strategies to reduce the emission of N_2O. Future research should first give better insight into the variability of N_2O emissions over prolonged times in order to develop proper sampling protocols for emissions of N_2O from treatment plants. A better understanding of the processes leading to nitrite formation in treatment plants can lead to mitigation strategies for N_2O emissions. Besides activated sludge systems, biofilm processes should be monitored for potential emissions. A better insight in N_2O production in natural systems can improve decisions on implementing full nitrogen removal as a mitigation strategy for N_2O emissions from treatment systems as well as natural systems.

6.3 CH_4 EMISSION

6.3.1 Sewers

The measurements performed in Australia and the USA show that substantial amounts of methane can be formed and emitted from sewer systems. In addition to this, substantial amounts of methane were measured from the inlet works at the WWTPs in the Netherlands. With this knowledge, the contribution of the sewer system cannot be neglected as assumed in the current IPCC guidelines (IPCC 2006a). However, to be able to make a good estimate of the contribution of sewers to the total CH_4 emission from wastewater collection and treatment systems more data are required and a good strategy to measure methane loads is necessary (refer to section 6.1.2). The data collected to date are far from adequate for the development of reliable accounting guidelines, and for the development, calibration and validation of CH_4 emission models. It is of utmost importance that such data are collected from sewer networks (rising mains, and gravity sewers) around the world under different climate conditions, and from networks collecting and conveying various types of wastewaters. Both liquid and gas phase data are urgently needed.

The results from the lift stations (fed mostly by gravity sewers) in the USA (Figure 5.8) showed that the concentration of methane in the gas phase was in almost all cases (80%) higher in summer than in winter. This indicates that temperature is an important parameter determining the formation of methane in sewers. This might also explain relative high methane levels in Australia, although this could also result from differences in the composition of wastewater.

6.3.2 Mitigation strategies

Several strategies used for sulphide control in sewers have been found to reduce methane formation (Figure 5.9–5.11). However these strategies may not be the most cost-effective methods for methane control. Given the high sensitivity of methanogens to environmental changes, more cost-effective strategies may be developed. The addition of nitrite is a very promising technology in this respect, but might easily lead to N_2O formation.

6.3.3 WWTPs

Based on the measurement in the Netherlands methane emission at a WWTP seems to originate from two major sources. First the sewer seems to be a major source as a high methane emission was observed from the inlet works and the aeration tanks. The second major source seems to be related to the sludge digester as high emissions of methane were observed from sludge handling facilities related to the sludge digester. These are the facilities that store the digested sludge before and after the dewatering facilities. In these facilities the methanation is still progressing, and methane dissolved in the liquid will be stripped. The methane that is formed inside the sludge digestion tank is combusted in combined heat power equipment, and it is thought that the leakage of CH_4 is minor.

For the storage facilities technological and technical adaptations will have to be developed in order to minimise the emissions of methane. Also the potential to have the methane biologically oxidised before it is stripped will have to be investigated.

From the results at the WWTP it is shown that process units that are known for their sulphur related odours (inlet works, sludge digesters) also emit methane.

6.4 TOTAL CARBON FOOTPRINT WWTP

As a first indication on the possible contribution of N_2O and CH_4 emission to the total carbon footprint of a WWTP, the result in the Netherlands could be used as an example.

In the case studies in the Netherlands, the specific emissions of N_2O and CH_4 were determined at the same time. Together with the data on the related consumption of electricity and natural gas, it was possible to calculate a carbon footprint of three WWTPs. To determine the carbon footprint, all sources were converted to CO_2 equivalents[1]. The results in the Netherlands indicated that the emission of CH_4 and N_2O can significantly contribute to the total carbon footprint of a WWTP. This contribution can vary from 2% to almost 90% of the carbon footprint under extreme conditions for N_2O and 5–40% for CH_4. One should be aware that these numbers are specific for the Netherlands. In any other country, these numbers can differ greatly as there exist a great variation in the way wastewater and sludge is handled as well as the specific composition of the energy mix used. Furthermore these numbers can significantly differ depending on how the boundaries are set around the analysis. In case of the analysis performed for the three Dutch WWTPs the contribution of e.g. chemical use, and sludge incineration were not accounted for.

[1]It should be noted that the conversion numbers are country specific and do depend on the used energy mix (i.e. brown coal versus wind or solar energy), which is of influence on the total carbon footprint of a WWTP.

Chapter 7

Conclusions and future research

7.1 CONCLUSIONS

7.1.1 N_2O emission

Based on the research presented in this report the following conclusions can be drawn:

- The emission of N_2O is highly variable. For this reason the use of generic emission factors to estimate the emission from a specific WWTP is inadequate.
- To determine the emission of a specific WWTP, measurements are required. These measurements should be performed with online monitors and over the operational range of the WWTP.
- Emission of N_2O originates mainly from nitrification.
- Accumulation of nitrite leads to the formation of N_2O in mainly in aerobic zones.
- High NH_4 concentrations can lead to the emission of N_2O if nitrification occurs.
- The risk level of N_2O emission can be determined based on the following risk matrix:

Table 7.1 Risk matrix to determine risk level of N_2O emission.

Risk on N_2O	High	Medium	Low
Parameter			
Effluent total nitrogen (mg/l)	>10	5–10	<5
Range in N-concentration in plant	H	M	L
Load variations (daily)	H	M	L
Maximum NO_2 concentration (mgN/l) anywhere in plant	>0.5*	0.2–0.5	0.2

*Risk does not increase at higher NO_2 concentrations.

- A good effluent quality (TN < 5 mgN/l) goes hand in hand with a low risk of N_2O emission.
- Non-BNR systems with unintentional nitrification will have a high risk of N_2O emission.

7.1.2 CH_4 emission

Based on the research presented in this report the following conclusions can be drawn:

- Formation and emission of CH_4 from sewers can be substantial.
- Sulphur related odours are good indicators for methane formation.
- Odour mitigation strategies in sewers likely also support a reduced methane formation.
- Methane emission from WWTPs mainly originates from sewers and sludge handling.

7.1.3 Total carbon footprint

- Emissions of CH_4 and N_2O can significantly contribute to the total greenhouse gas footprint of WWTPs as was seen from the results obtained in the Netherlands, but this can be different in other countries.

7.2 FUTURE RESEARCH

Based on the outcomes of the research valuable knowledge is gained to estimate and reduce the emission of N_2O and CH_4 from wastewater collection and treatment systems. Future applied research should focus on:

N_2O

- Variability of N_2O emission such that proper sampling and monitoring programmes can be developed.
- Mitigation strategies (including insight in the relative contribution of autotrophic and heterotrophic processes to N_2O generation) to reduce emission via process design and control;
- Evaluate emission from biofilm based processes;
- Evaluate emission from various receiving aquatic environments.

CH_4

- Development of a strategy to measure loads emitted from sewers;
- Build a database for emission measurements from sewers;
- Cost effective mitigation strategies[1]
- Emission from sludge treatment lagoons.

[1]The addition of nitrite seemed to be a strategy that could reduce the formation of methane, however, given that nitrite is one of the factors that correlate with N_2O production, the impact of upstream nitrite addition and dosages on the potential on downstream N_2O emissions needs to be determined as well.

Chapter 8
Acknowledgements

8.1 AUSTRALIA METHANE RESEARCH

The following people from the Advanced Water Management Centre at The University of Queensland are thankfully acknowledged for conducting the original research reported in this document on the measurement, understanding, modeling and mitigation of methane formation in sewers: David de Haas, Jeff Foley, Albert Guisasola, Oriol Gutierrez, Guangming Jiang, Jurg Keller, Paul Lant, Janani Mohanakrishnan, Keshab Sharma, Gatut Sudarjanto and Lishan Zhang.

The Australian Research Council (ARC), the Water Services Association of Australia (WSAA) and numerous members of the Australian water industry provided financial support to the research presented in this report through ARC projects LP0454182, ARC LP0882016 (http://www.score.org.au/) and the WSAA project Direct Methane and Nitrous Oxide emissions from full-scale wastewater treatment systems.

Dr Trisha Ibrahim from South East Water Limited (SEWL) kindly provided gas-phase methane data measured at one of the sewer lines operated by SEWL. The discussion with Trisha about these results has been very highly valuable.

Mr Shaun Corrie and Mr Ian Johnston from Allconnex Water (formerly Gold Coast Water) assisted with the collection of dissolved methane data from UC09 and CO16. They also provided on-line gas phase methane data measured at CO16.

8.2 AUSTRALIA NITROUS OXIDE RESEARCH

This work was funded by the Water Services Association of Australia and The University of Queensland. In-kind support was also provided by United Water International (Adelaide), and the participating water authorities.

The members of the project team for this study were:

Principal authors:	Jeff Foley (PhD student, UQ)
	Professor Paul Lant (UQ)
Fieldwork assistants:	Bill Fuller (Engineer, United Water International)

	Peter Muller (Undergraduate engineer, GHD)
	Con Foley (volunteer)
	Lyndon Clark and Debbie Chua (Water Corporation)
	Lan Johnson (Gold Coast Water)
Analytical and modelling support:	Dr Ursula Werner (UQ)
	Shihu Hu (PhD student, UQ)
	Dr David de Haas (UQ, GHD)
	Dr Keshab Sharma (UQ)
	Dr Albert Guisasola (Universitat Autònoma de Barcelona)
	Dr Beatrice Keller-Lehmann and Jianguang Li (UQ)
	Peng Su and Associate Professor Wei Zhang (Flinders University)
Technical review and stewardship:	Professor Zhiguo Yuan (UQ)
	Professor Jurg Keller (UQ)
	Dr David de Haas (UQ, GHD)
	Dr Paul Rasmussen (United Water International)
	Hamish Reid (South East Water)
	Michael Greg and Daniel Starrenburg (Brisbane City Council)
	Greg Allen (Sydney Water)
	Keith Cadee (Water Corporation)
	Adam Lovell (Science and Sustainability Manager, WSAA)

The authors also thank the plant owners and operators for their cooperation:

Brisbane City Council	Keith Barr, Vijay Jayaraman, Kenny Liew and Grahame Simpson
Fosters Brewery	Charlie Foxall
Gold Coast Water	Shaun Corrie, Ian Johnson and Kelly O'Halloran
Melbourne Water	Kim Daire, Peter Gall, Trevor Gulovsen, Suelin Haynes and Erik Ligtermoet
SA Water/United Water	Tim Kelly, Bridget Kingham, Ian MacKenzie and Dr Paul Rasmussen
Sydney Water	Heri Bustamante, Steven Lin, Mark Ognibene, Dammika Vitanage, Wanxin Wang and Wasantha Wicks
Toowoomba Regional Council	Alan Kleinschmidt and John Paulger
Western Australian Water Corporation	Claire Camplin, Lyndon Clark, Margaret Domurad, Kay Hyde, Pardeep Kumar, Gabrielle O'Dwyer and Leah Rheinberger

8.3 UNITED STATES OF AMERICA NITROUS OXIDE RESEARCH

This work was funded by the Water Environment Research Foundation under Project U4R07 and the Climate Change program. The members of the project team for this study were:

Principal authors:	Kartik Chandran, Ph.D. (Columbia University)
Field researchers:	Joon Ho Ahn, Ph.D. (Columbia University) Sungpyo Kim, Ph.D. (Columbia University) Hongkeun Park, Ph.D. (Columbia University) Brian Rahm, Ph.D (Columbia University)
Laboratorium researchers	Ran Yu, Ph.D. (Columbia University) Huijie Lu, Ph.D. (Columbia University)
Technical Review and Direction:	Greg Adams, P.E. (Los Angeles County Sanitary District) Eugenio Geraldo, Ph.D. (American Water) William Hiatt, Ph.D., P.E. Sudhir Murthy, Ph.D., P.E. (DC Water) Denny Parker, Ph.D. P.E. (Brown and Caldwell) Peter Strom, Ph. D. (Rutgers University) Jeff Weiss, Ph.D. (Los Angeles County Sanitary District)

8.4 UNITED STATES OF AMERICA METHANE RESEARCH

This work was funded by the Water Environment Research Foundation under Project U2R08 and the Climate Change program. The members of the project team for this study were:

Principal authors:	John Willis, P.E. (Brown and Caldwell)
Field assistants:	Anup Shah (Brown and Caldwell)
Technical Review and Direction:	Bob Forbes, PE (CH2M-Hill) Eugenio Geraldo, Ph.D., (American Water) Catherine O'Connor, (Metropolitan Water Resources District of Greater Chicago) Eliza Jane Whitman, PE (Parsons) Patrick Wootton, (GE)

8.5 RESEARCH PERFORMED IN THE NETHERLANDS

The authors acknowledge the contribution of Kyra Schipper, Pieter Vijfhuizen in the collection of field data. The Steering Committee of the project, Alex Sengers, Klaas Appeldoorn, Frank Brandse and Hardy Temmink are thanked for their constructive input during the project. The plant operators at Kralingseveer, Kortenoord and Papendrecht WWTP are also acknowleged for their hospitality and co-operation.

Chapter 9
References

Ahn, J.-H., Kim, S., Pagilla, K., Katehis, D. and Chandran, K., 2009, Spatial and temporal variability in N_2O generation and emission from wastewater treatment plants. Nutrient Removal Conference. Washington D.C., Water Environment Federation

Andrews, J. Chambers, B., Davey, A., Galletti, S., Hobson, J., Hunt, D., Thorman, R. and Walker, I., 2009, Carbon Accounting in the Water Industry; non-CO_2 Emissions, UKWIR, London

Burgess, J. E., Colliver, B. B., Stuetz, R. M. and Stephenson, T., 2002, Dinitrogen oxide production by a mixed culture of nitrifying bacteria during ammonia shock loading and aeration failure, Journal of Industrial Microbiology & Biotechnology 29, pp. 309–313.

Casey, T. G., Wentzel, M. C. and Ekama, G. A., 1999a, Filamentous organism bulking in nutrient removal activated sludge systems – Paper 10: Metabolic behaviour of heterotrophic facultative aerobic organisms under aerated/unaerated conditions. Water Sanitation 25 (4), 425–442.

Casey, T. G., Wentzel, M. C. and Ekama, G. A., 1999b, Filamentous organism bulking in nutrient removal activated sludge systems Paper 9: Review of biochemistry of heterotrophic respiratory metabolism. Water Sanitation 25 (4), 409–424.

Chandran, K., 2010, Greenhouse nitrogen emission from wastewater treatment operations, WERF report U4R07a.

Czepiel, P. M., Crill, P. M. and Harriss, R. C., 1993, Methane emissions from municipal wastewater treatment processes, Environmental Science Technology, 27 pp. 2472–2477.

Foley, J. and Lant, P., 2008. Fugitive Greenhouse Gas Emissions from Wastewater Systems: WSAA Literature Review No.01. Water Services Association of Australia, Melbourne.

Foley, J. and Lant, P., 2009, Direct Methane and Nitrous oxide emissions from full-scale wastewater treatment systems, Occasional paper No.24, Water Service Association of Australia.

Gray, N. D., Miskin, I. P., Kornilova, O., Curtis, T. P. and Head, I. M., 2002, Occurrence and activity of *Archea* in aerated activated sludge wastewater treatment plants, Environmental Microbiology 4 (3), pp. 158–168.

Guisasola, A., Haas, de D., Keller, J. and Yuan, Z., 2008, Methane formation in sewer systems, Water Research (42) pp. 1421–1430.

Gutierrez, O., Park, D., Sharma, K. R. and Yuan, Z., 2009, Effects of long-term pH elevation on the sulphate-reducing and methanogenic activities of anaerobic sewer biofilms. Water Research 43 (9), 2549–2557

GWRC, 2011, N_2O and CH_4 Emission from Wastewater Collection and Treatment Systems, Technical report, London, United Kingdom, ISBN 978–90–77622–24–7.

Hvitved-Jacobsen, T., 2002, Sewer processes: microbial and chemical process engineering of sewer networks. CRC Press, USA.

IPCC, 2006a. Ch.6 - Wastewater Treatment and Discharge. in H. S. Eggleston, L. Buendia, K. Miwa, T. Ngara, and K. Tanabe, editors. 2006, IPCC Guidelines for National Greenhouse Gas Inventories, Prepared by the National Greenhouse Gas Inventories Programme, vol.5 - Waste. IGES, Japan.

IPCC, 2006b. Vol.1 - General Guidance and Reporting. in H. S. Eggleston, L. Buendia, K. Miwa, T. Ngara, and K. Tanabe, editors. 2006, IPCC Guidelines for National Greenhouse Gas Inventories, Prepared by the National Greenhouse Gas Inventories Programme, IGES, Japan.

IPCC (2007), Fourth Assessment Report (AR4), Climate Change 2007: Synthesis report.

Jiang, G., Gutierrez, O., Sharma, K. R. and Yuan, Z., 2010, Effects of nitrite concentration and exposure time on sulfide and methane production in sewer systems. Water Research, 44 (14), 4241–4251.

Kampschreur, M. J., Temmink, H., Kleerebezem, R., Jetten, M. S. and van Loosdrecht, M. C., 2009, Nitrous oxide emission during wastewater treatment, Water Research 43 (17), pp. 4093–4103.

Kampschreur, M. J., 2010, Dynamics of nitric oxide and nitrous oxide emission during nitrogen conversion processes, PhD thesis Delft University of Technology.

Lens, P. N., De Poorter, M.-P., Cronenberg, C. C. and Verstraete, W. H., 1995, Sulfate reducing and methane producing bacteria in aerobic wastewater treatment systems, Water research (29–3) pp. 871–880.

Tata, P., Witherspoon, J. and Lue-Hing, C. (Eds.), 2003, VOC Emissions from Wastewater Treatment Plants, Boca Raton, FL, Lewis Publishers.

Tallec, G., Garnier, J., Billen, G. and Gousailles, M., 2006. Nitrous oxide emissions from secondary activated sludge in nitrifying conditions of urban wastewater treatment plants: Effect of oxygenation level. Water Research 40 (15), 2972–2980.

Tchobanoglous G & F L Burton, 2003, Wastewater engineering – Treatment and reuse Metcalf & Eddy, Inc. 4th edition 2003. MacGraw-Hill, New York 2003

Voorthuizen van, E. M., van Leusden, M., Visser, A., Kruit, J., Kampschreur, M., Dongen van, U. and Loosdrecht van, M., 2010, Emissies van broeikasgassen van rwzi (in Dutch, summary in English), STOWA report 2010–08.

VROM, 2008, Protocol 8136 Afvalwater, t.b.v. NIR 2008, uitgave maart 2008; 6B CH_4 en N_2O uit afvalwater (in Dutch).

UNFCCC, 2007, Uniting on climate. United Nations Framework Convention on Climate Change, Bonn.

Yu, R., Kampschreur, M. J., Loosdrecht, M. C. M. V. and Chandran, K., 2010, Mechanisms and Specific Directionality of Autotrophic Nitrous Oxide and Nitric Oxide Generation during Transient Anoxia. Environmental Science Technology, 44, 1313–1319.

Zhang L., Keller, J. and Yuan, Z., 2009, Inhibition of sulphate-reducing and methanogenic activities of anaerobic sewer biofilms by ferric iron dosing. Water Research 43(17), 4123–4132.

Appendix 1.1
Sample collection with gas hoods

SAMPLE COLLECTION FRANCE

The way the samples were taken for N_2O analysis depended on whether the surface was aerated or non-aerated. Both protocols are described below.

Sample collection from aerated surfaces

The square wooden sampling box (1m * 1m) is used in aerated conditions (see Figure A1.1.1). A pipe connected to a gas mass counter allows the flow of air supplied by the aeration system – 'real' Q – to be collected and quantified (in m^3/h). A spur from this pipe allows the sample of air to be directed to the gas analyzer.

Figure A1.1.1 Sampling box for aerated surfaces.

Sample collection from non-aerated surfaces

In case of non-aerated surfaces the sampling system used was the Odoflux dynamic flow chamber (see Figure A1.1.2).

Figure A1.1.2 Sampling box for non-aerated surfaces.

The flow chambers covered an area to be studied as hermetically as possible in order to isolate the surface from external conditions. The gases emitted by the isolated surface are collected by a vector gas which is injected into the chamber (Q1). The mixture of vector gas and gaseous effluent is then collected for analysis. The flow chamber used consists of an acrylic resin cylinder capped by a hemisphere, also of acrylic resin. The air-tightness of these two parts is ensured by polypropylene screws. The sweep air supply is fed by a pipe coiled against the wall and pierced with holes (flow Q1). The air sample is taken at the top of the hemisphere by means of a probe which has several holes along its length in order to ensure homogeneous sampling. The air is sucked in at flow Q2, of which 1l/min is directed to the analyser; the other part (Q2–1l/min) is released into the atmosphere.

To respect 'pseudo-isokinetic' conditions, the applied flow Q1 must be equal to the flow sucked in Q2. The protocol developed in the framework of this project for measuring GHG recommends a relatively weak flow (Q1 = Q2) of the order of 3 to 10l/min; this flow thus ensures 'static chamber' conditions in which the surface of the liquid is not disturbed and the surface emissions are very slightly diluted.

ANALYSIS N_2O IN LIQUID PHASE

The protocol for the analysis of dissolved N^2O presented here was drawn up by the Rennes CEMAGREF. The water samples are taken and conditioned in 330ml flasks, sealed hermetically with a septum. The sample conditioning stages are detailed below

- Weigh 100g of water sample in a 330ml glass flask.
- Add 1 drop of sulphuric acid and close the flasks hermetically (stopper + septum).

- Place the samples in a container with warm water at 80°C for 1hr, the dissolved N_2O is then released into the head space.
- The gaseous fraction in the head space is then sampled using a gas syringe and injected into three 3ml pill-boxes which have been previously conditioned in a vacuum;
- The analysis of the dissolved N_2O is done by gas chromatography – electron capture detector (GC/ECD) by the Rennes CEMAGREF, in France.

SAMPLE COLLECTION UNITED STATES OF AMERICA

Sampling procedures-headspace gas measurement

The overall procedure for measuring N_2O, NO and NO_2 fluxes from the head-space of activated sludge tanks involves a variant of the EPA/600/8-86/008 and the South Coast Air Quality Management District (SCAQMD) tracer methods. This variant was developed to measure those sources that have a relatively high surface flux rate when compared to diffusion (for instance, spilled oil containment).

Commercially available replicas of the US EPA surface emission isolation flux chamber (SEIFC, Figure A1.1.3) were used to measure gaseous N fluxes from activated sludge reactors.

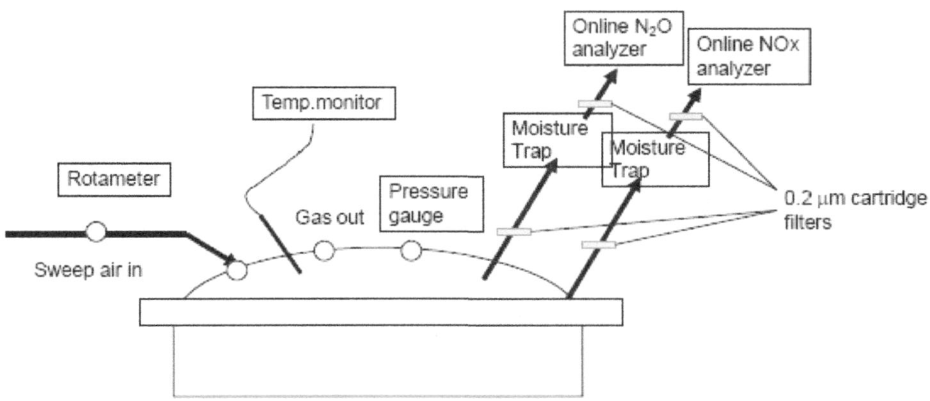

Figure A1.1.3 Schematic of Flux-chamber Employed for N_2O measurement.

The SEIFC consists of a floating enclosed space from which exhaust gas is collected in a real-time or discrete fashion. Since the surface area under the SEIFC can be measured, the specific flux of the gaseous compound of interest can be indirectly determined. The SEIFC 'floats' on the activated sludge tank surface and several replicate measurements can be taken at different locations in a single tank as well as from different tanks (nitrification, denitrification) along a treatment train.

The SEIFC is also equipped with mixing (physical mixer or via sweep gas circulation) to ensure adequate gas mixing and in some cases, an online temperature probe. The SEIFC is currently one of the few devices accepted by the USEPA for measuring gaseous fluxes (Tata et al., 2003) and as such it was employed for this study.

Appendix 1.2

Abbreviations and glossary

Abbreviation / term	
A/O and A^2/O	Type of WWTP configuration for the biological removal of phosphate[1]
ADEME	French Environment and Energy Management Agency
Bardenpho	Type of WWTP configuration for the biological removal of phosphate[1]. Can be designed with different amount of stages.
BNR	Biological Nitrogen Removal
CH_4	Methane
CO_2	Carbon Dioxide
COD	Chemical Oxygen Demand
DO	Dissolved Oxygen
EPA	Environmental Protection Agency
FB	Fermentative Bacteria
$Fe^{2/3+}$	Iron
FID	Flame Ionization Detector
GC	Gas Chromatograph
GF_{WWTP}	Generation factor (N_2O) for whole WWTP
GHG	Greenhouse Gas
H_2S	Hydrogen Sulphide
HRT	Hydraulic Retention Time
IPCC	Intergovernmental Panel on Climate Change
Johannesburg	Type of WWTP configuration[1]
K_La	Mass transfer coefficient (liquid → gas)

LCC – PS	Lower Crooked Creek Pump Station
MBR	Membrane Bioreactor
MLE	Modified Ludzack-Ettinger (i.e. a configuration type)[1]
NH_4(-N)	Ammonium
NKj	Nitrogen Kjeldal
NGERS	National Greenhouse and Energy Reporting System
NO_2 (-N)	Nitrite
NO_3 (-N)	Nitrate
N_2O	Nitrous Oxide
SBR	Sequenced Batch Reactor
SEIFC	Surface emission isolation flux chamber
SRB	Sulphate Reducing Bacteria
SRT	Sludge retention time
TKN	Total Kjeldahl Nitrogen
VFA	Volatile Fatty Acids
USEPA	United States Environmental Protection Agency
WWTP	Wastewater Treatment Plant

[1]For more information see: Tchobanoglous, G., 2003.

Appendix 1.3

DeKalb County's collection system

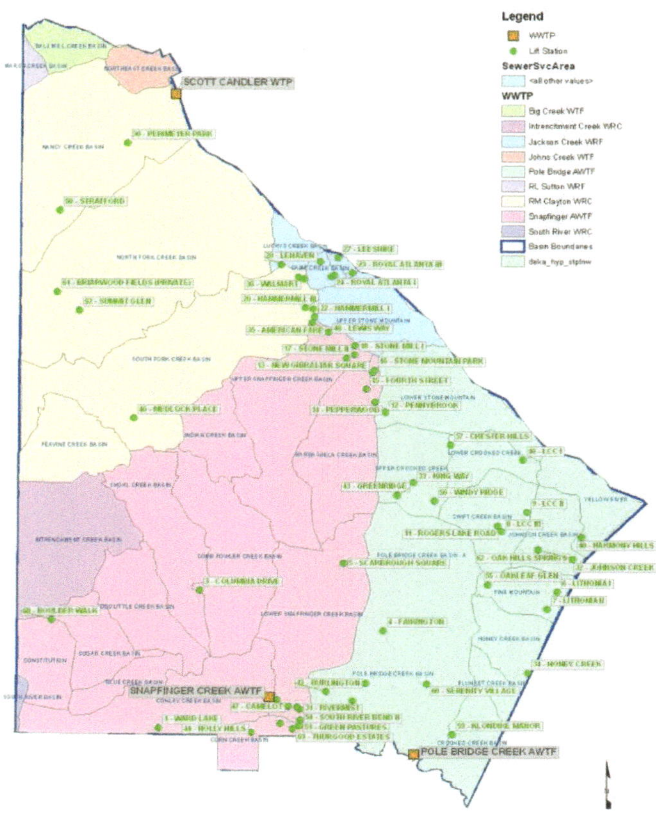

Figure A1.3.1 Map of DeKalb County's collection system.

Appendix 1.4

Characteristics WWTPs investigated in N_2O research

Country	Type of process	Sludge load[1]	Temperature	SRT (appr.)	Effluent TN[1]	Process description
		(kg TKN_{in}·kg d.w.$^{-1}$·d^{-1})	(°C)	(d)	(mg/l)	
Australia	1. Ox.Ditch (4 rounds)[2]	0.043 0.042 0.03 0.042	23	13	3	Inlet works, anaerobic contact tank, 2 × extended aeration oxidation ditches (in parallel) with diffused aeration, secondary sedimentation; mechanical sludge thickening and dewatering.
	2. Johannesburg (3 rounds)	0.012 0.01 0.013	21	20	5	Inlet works, 2 × extended aeration Johannesburg bioreactors (in parallel) with submerged aspirating OKI™ aerators, secondary sedimentation, sludge thickening, aerobic digestion, mechanical and solar dewatering.
	3. SBR (3 rounds)	0.021 0.017 0.021	20	16	15	Inlet works, primary sedimentation, sequencing batch reactor (4 compartments) with diffused aeration and bio-selector zone, sludge thickening, anaerobic digestion and mechanical dewatering.
	4. MLE-1 (1 round)	0.031	21	13	12	Inlet works, primary sedimentation, 11 × covered Modified Ludzack Ettinger (MLE) bioreactors (in parallel) with diffused aeration, secondary sedimentation, sludge thickening, mechanical dewatering and lime stabilisation.
	5. MLE-2 (4 rounds)	0.06 0.065 0.072 0.07	20	8	11	Inlet works, primary sedimentation, 2 × Integrated Fixed Film Activated Sludge bioreactors + 1 × MLE bioreactor (in parallel) with diffused aeration, secondary sedimentation, sludge thickening and anaerobic digestion.
	6. MLE-3 (3 rounds)	0.01 0.01 0.012	18	15	13	Anaerobic lagoon, 1 × MLE bioreactor with diffused aeration, secondary sedimentation, maturation lagoons, and sludge wasting to an aerated facultative lagoon.
	7. A^2/O (2 rounds)	0.009 0.01	21	14	3	Inlet works, 2 × parallel trains: 1) Primary sedimentation, 4-stage Bardenpho bioreactor with

						diffused aeration, secondary sedimentation; 2) Pre-fermenter, four-stage bioreactor (similar to A^2/O configuration) with diffused aeration and supplemental COD dosing by primary sludge from Train 1, secondary sedimentation, tertiary filtration, sludge thickening, aerobic digestion, sludge lagoon and mechanical dewatering.
France	1. A^2/O	n.a.	n.a.	n.a.	n.a.	Three-zones completely Stirred Tank Reactor. (anaerobic- anoxic- aerobic with sequenced aeration)
	2. A/O	n.a.	n.a.	n.a.	n.a.	Plug flow process (anoxic –aerobic)
	3. A^2/O	n.a.	n.a.	n.a.	n.a.	Three-zone completely Stirred Tank Reactor (anaerobic- anoxic- aerobic with sequenced aeration)
	4. ?? / MBR					Activated sludge with sequenced aeration followed by MBR
USA	1. Separate stage BNR (2 rounds)[2]	n.a.	15 23			The low-rate separate-stage nitrification denitrification process at this WWTP was sampled. The process was configured as a sequence of five reactors in series. The influent to this process consisted of the clarified effluent from an upstream high-rate process, mainly engaged organic carbon removal. The influent was fed in a step-feed fashion to the first two aerobic zones. The last three zones of this process were non-aerated and the second non-aerated zone received methanol to promote denitrification. The effluent channel of this process was aerated prior to secondary clarification.
	2. Four-stage Bardenpho (2 rounds)	n.a.	14 23			The four-stage Bardenpho process consisted of pre-denitrification (without external carbon addition) followed by a primary aerated zone. The effluent of the primary aerated zone was internally recycled to the anoxic zone. Following the primary aerated zone was a de-oxygenation zone to scavenge dissolved oxygen, prior to methanol addition for enhanced denitrification. The final zone in this process was aerated primarily for stripping off the dinitrogen gas produced during denitrification, prior to secondary clarification

(*Continued*)

Country	Type of process	Sludge load[1]	Temperature	SRT (appr.)	Effluent TN[1]	Process description
	3. Step-feed BNR 1 (2 rounds)	n.a.	19 25			The four-pass step-feed BNR process sampled consisted of pre-anoxic zones comprising about 1/3 of the pass volume followed by aerated zones. The transition zone between each pass was non-aerated to facilitate deoxygenation. The approximate influent flow split was10%–40%–30%–20% to passes A, B, C and D, respectively. Pass A also received pre-settled anaerobic digestion centrate, which constituted approximately 30% of the influent TKN load to the process. Return activated sludge was also fed to Pass A.
	4. Step-feed non BNR (2 rounds)	n.a.	17 26			The step-feed non-BNR process sampled was configured and operated in four-pass step-aeration mode. The process was completely covered primarily for odor control. The headspace off-gases were consolidated and fed to a biofilter. The approximate influent flow split was10%–40%–30%–20% to passes A, B, C and D, respectively. Return activated sludge was fed to Pass A.
	5. Separate centrate (2 rounds)	n.a.	30 34			The separate centrate treatment process was operated to process pre-settled anaerobic digestion centrate and partially convert the influent NH_4^+-N to NO_2^--N. The separate centrate treatment process was operated in plug flow mode. Effluent from the separate centrate tank was fed to the overall plant return activated sludge line for possible bioaugmentation with primarily ammonia oxidizing bacteria (AOB) and for nitrogen removal via the short-cut nitrite pathway.
	6. Plug flow 1 (2 rounds)	n.a.	11 23			The first plug-flow process sampled was designed and operated primarily for organic carbon removal and nitrification and did not have dedicated anoxic zones or external organic carbon addition. The process was configured in four-pass mode.
	7. Plug flow 2	n.a.	11			The second plug-flow process sampled was also

(2 rounds)		22	designed and operated for organic carbon removal and nitrification and did not have dedicated anoxic zones or external organic carbon addition. The process was configured in two-pass mode.
8. MLE-1 (1 round)	n.a.	26	The first modified Lutzack Ettinger (MLE) process sampled was originally designed for operation in enhanced biological phosphorous removal mode, but subsequently operated in MLE mode. The process consisted of pre-denitrification without external organic carbon addition. Biomass from zone 6, which was aerated, was internally recycled to anoxic zone 2. Return activated sludge was fed to anoxic zone 1.
9. MLE-2 (1 round)	n.a.	26	The second modified Lutzack Ettinger (MLE) process sampled was also originally designed for operation in enhanced biological phosphorous removal mode, but subsequently operated in MLE mode. The process consisted of pre-denitrification without external organic carbon addition. Biomass from zone 6, which was aerated, was internally recycled to anoxic zone 2. Return activated sludge was fed to anoxic zone 1.
10. Step-feed BNR 2 (1 round)	n.a.	29	The second step-feed process sampled was configured in four-pass mode. Each pass consisted of pre-anoxic zones comprising 1/3 of the pass volume followed by aerobic zones. The approximate influent flow split was50%–30%–20%–0% to passes A, B, C and D, respectively. The anoxic zones were mixed via low intensity pulse aeration. The return activated sludge was fed to Pass A.
11. Oxidation ditch (1 round)	n.a.	19	The oxidation ditch process was operated to achieve simultaneous nitrification and denitrification by operation are uniformly low aeration intensities and dissolved oxygen concentrations. The influent flow to the process was fed to the inner loop and was mixed and circulated using surface mixers. No external organic carbon was added to enhance denitrification. Return activated sludge was fed to the inner loop of the process.

(*Continued*)

Country	Type of process	Sludge load[1]	Temperature	SRT (appr.)	Effluent TN[1]	Process description
	12. Step-feed BNR 3 (1 round)	n.a.	24			The third four-pass step-feed BNR process sampled consisted of pre-anoxic zones comprising about 1/3 of the pass volume followed by aerated zones. The approximate influent flow split was 33.3%–33.3%–33.3%–0% to passes A, B, C and D, respectively. Pass A also received pre-settled anaerobic digestion centrate, which constituted approximately 40% of the influent TKN load to the process. Return activated sludge was also fed to Pass A. The reactors of this process were also covered and thus only composite measurements of the overall headspace could be performed.
Netherlands	1. Carrousel N&P removal; Papendrecht	0.009	20	23	4	Upfront the carrousel with simultaneously nitrification/denitrification a selector and anaerobic reactor was situated.
	2. Carrousel N&P removal; Kortenoord	0.012	19	n.a.	2.2	Upfront the carrousel with simultaneously nitrification/denitrification a selector and anaerobic reactor was situated.
	3. BNR; Kralingseveer (2 rounds)[3]	0.024 (fall) 0.036 (winter)	18 9	20	8.3 (fall) 10.4 (winter)	In this BNR the following reactor sequence was present: selector, plug flow denitrification zone, plug flow nitrification with recycle to denitrification zone, two parallel carrousels with simultaneous nitrification/denitrification.

[1]Based on actual data during measurement campaigns. [2]Explanation rounds see table 2. [3]Measured once in fall and once in winter.

Appendix 1.5

Research area methane research Australia

Figure A1.5.1 Aerial photograph of the CO16 rising main. Samples were collected from the CO16 Pump Station wet well (0 m), Sample point 1 (at 500 m), Sample point 2 (at 1100 m) and Sample point 3 (at 1900 m).

Part II
Technical Report

Chapter 10
Introduction

10.1 BACKGROUND

In a world where there is a growing awareness of the possible effects of human activities on climate change, there is a need to identify the emission of greenhouse gases (GHG) from wastewater treatment plants (WWTPs)[1]. As a result of this growing awareness, governments started to implement regulations that require water authorities to report their GHG emissions. With these developments there exists a strong need for adequate insight into the emissions of N_2O and CH_4. With this insight water authorities would be able to estimate and finally reduce their emissions. At the time little information was available on the formation of GHG, and the emission factors used by the IPCC are based on limited data. The limits of available data became the driver to start extensive field studies in Australia, France, the United States of America and the Netherlands with the objective to fill the knowledge gaps needed to estimate and reduce the emission of N_2O and CH_4 from wastewater collection and treatment systems.

The research programs were performed by partners[2] of the GWRC members WERF (United States of America), WSAA (Australia), CIRSEE-Suez (France) and STOWA (the Netherlands).

10.2 OBJECTIVES

The overall objectives of the different research programs[3] were:

- Define the origin of N_2O emission.
- Understand the formation processes of N_2O.
- Identify the level of CH_4 emissions from wastewater collection and treatment systems.
- Evaluate the use of generic emission factors to estimate the emission of N_2O from individual plants.

[1]The greenhouse gases associated with the activities at WWTPs are CO_2, CH_4 and N_2O. Of these gases, N_2O is the most important as it has a 300-fold stronger effect than CO_2. CH_4 is less strong than N_2O but still has a 25-fold stronger effect than CO_2. Nitrous oxide (N_2O) can be formed during the conversion of nitrogenous compounds in wastewater; methane may be emitted in the sewer system and during sludge handling. The emission of CO_2 from the biological treatment is part of short cycle (or biogenic) CO_2 and does not contribute to thecarbon footprint. However, some carbon in wastewater may originate from fossil fuel.

[2]Partners were: Columbia University, USA; Brown and Caldwell, USA; The University of Queensland, Australia; Delft University of Technology, the Netherlands, Royal Haskoning, the Netherlands.

[3]In the technical report (GWRC, 2011) that accompanies this State of the Art Report the objectives of the individual partners are mentioned.

10.3 ACTIVITIES WITHIN THE GLOBAL WATER RESEARCH FRAMEWORK

The topic of N_2O emissions from wastewater treatment facilities is part of the research area Climate Change of the joint research agenda of the Global Water Research Coalition (GWRC). STOWA took the lead to develop and coordinate this joint activity with support of the GWRC members Anjou Recherche, Eawag, CIRSEE, UKWIR, WERF, WRC and WSAA. Representatives of the involved members have met on several occasions making use of opportunities of planned conferences and workshops like the WWC in Vienna (September 2008), the GWRC workshop in Dübendorf (February 2009), and a meeting on the occasion of the SIWW/LET 2009 in Singapore (June 2009). An inventory of members research programs was performed by STOWA and detailed information on the ongoing efforts was discussed and protocols exchanged.

In August 2009 the GWRC N_2O website was launched and involved GWRC-members (and invited experts) can use the site to exchange information and comment results.

At present the members of the GWRC have either initiated or are planning to undertake research to measure the emission of N_2O from wastewater treatment facilities. An extensive research program was set up in Australia, the Netherlands and the United States of America to quantify the emission of N_2O and CH_4 from sewers and WWTPs. In these research programs there was a focus on the emission of N_2O, the emission of CH_4 was studied in less detail. The reason for this difference in focus is the fact that N_2O is a much stronger greenhouse gas than CH_4 and that little is known about the formation processes of N_2O in WWTPs.

10.4 ONGOING ACTIVITIES OUTSIDE GWRC

Besides the activities of the GWRC members, a new IWA Task Group will focus on the use of water quality and process models for minimizing wastewater utility greenhouse gas footprints. The main objectives of this group are:

- Understand the processes that are responsible for the major contributions to GHG emissions from WWTP and sewer systems.
- Incorporate this knowledge into mathematical models that can be embedded in system/plant-wide models allowing multi-criteria optimisation.

The World Bank, with partners, has financed an ongoing project at the Rio Frio wastewater treatment plant in Columbia to reduce CH_4 and N_2O emissions. The project had several objectives, including:

- Improvements in gas separation in the anaerobic reactors and during gas engines to result in additional abatement of CH_4 emissions.
- The reduction in N loads in the receiving waters will result in a corresponding reduction in N_2O.

10.5 BOUNDARIES REPORT

The research described in this report was the first extensive research on N_2O and CH_4 emission from wastewater collection and treatment systems. The main focus was to identify the level of emission, the variation therein and improve the knowledge on N_2O formation. Definition of mitigation strategies was outside the scope of most of the research as the knowledge on formation and orgin was too limited at the start of the research. For methane some mitigation strategies were investigated and are reported here.

10.6 OUTLINE REPORT

This report extensively describes the field and laboratory studies that have been performed in Australia, France, the United States of America and the Netherlands and presents a higher level of detail than the state of the art report on the topic (GWRC, 2011).

An extensive literature review is presented in chapter 11. The local regulations as they apply in countries participating in the GWRC report are presented in chapter 12. The individual objectives of the projects are presented in chapter 13. The methodology used by the individual countries is given in chapter 14. An overview of all the results is presented in chapter 15 and this is discussed in chapter 16. Finally the conclusions and recommendations for future research are presented in chapter 17. The following reports of the individual GWRC members were used:

- WERF: Chandran, K., 2010, Greenhouse nitrogen emission from wastewater treatment operations, WERF report U4R07a.
- WSAA: Foley, J. and Lant, P., 2009, Direct Methane and Nitrous oxide emissions from full-scale wastewater treatment systems, Occasional paper No.24, Water Service Association of Australia.
- STOWA: Voorthuizen van, E. M., van Leusden, M., Visser, A., Kruit, J., Kampschreur, M., Dongen van, U. and Loosdrecht van, M., 2010, Emissies van broeikasgassen van rwzi (in Dutch, summary in English), STOWA report 2010–08.

Chapter 11

Literature review

11.1 NON CO_2 GREENHOUSE GASES

The non CO_2 greenhouse gases that can be emitted from a domestic WWTP are nitrous oxide (N_2O) and methane (CH_4). The locations at a WWTP where these gases can be emitted are presented in Figure 11.1.

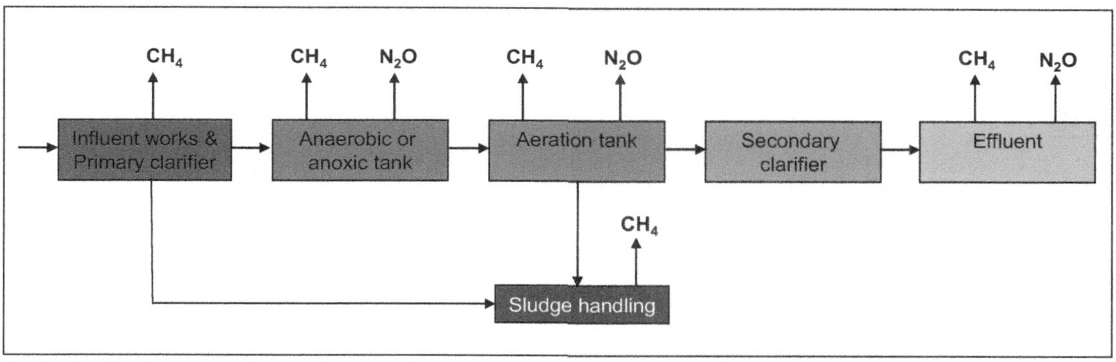

Figure 11.1 Schematic overview of a domestic WWTP and the locations where CH_4 and N_2O can be emitted.

Methane that is emitted from the influent works is most likely formed in the sewer system, as the retention time of the wastewater in the influent works is too short to form CH_4. Furthermore CH_4 formation will only occur where anaerobic or anoxic conditions prevail, as in the anaerobic or anoxic tank, but then only in the biofilms at the side of tanks, and at sludge handling sites. For this reason no CH_4 formation is expected in an aeration tank. Methane that is emitted here is formed earlier (in sewer or in sludge digester) and is stripped to the gas phase in the aeration tank. Formation and emission of N_2O can only occur under anoxic or aerobic conditions in the presence of nitrate (and carbon source) and ammonium. Nitrogen that is not converted leaves the WWTP via the effluent, which can lead to the emission of N_2O from surface water.

11.2 RELEVANT PROCESSES N_2O FORMATION

Nitrous oxide can be produced during the conversion of nitrogen in WWTPs. The processes involved are nitrification and denitrification. Besides N_2O formation by biological processes in activated sludge systems, there can be N_2O generation when e.g. biogas is burned at the WWTP for electricity production.

11.2.1 Nitrification

Nitrification is performed by three different groups of autotrophic microbes; ammonium-oxidizing bacteria (AOB) and ammonium-oxidizing archaea (AOA) that convert ammonia into nitrite, and nitrite-oxidizing bacteria (NOB) that convert nitrite into nitrate. The different steps involved in the nitrification are presented in Figure 11.2.

AOB / AOA			
	$NH_3 + O_2 + 2H^+ + 2e^-$	→	$NH_2OH + H_2O$
	$NH_2OH + H_2O$	→	$NO_2^- + 5H^+ + 4e$
	$0.5\ O_2 + 2H^+ + 2e^-$	→	H_2O
Total	$\mathbf{NH_3 + 1.5O_2}$	→	$\mathbf{NO_2^- + H^+ + H_2O}$
NOB			
	$NO_2^- + H_2O$	→	$^- + 2H^+ + 2H^+ + 2e^-$
	$0.5O_2 + 2H^+ + 2e^-$	→	H_2O
Total	$\mathbf{NO_2^- + 0.5O_2}$	→	-

Figure 11.2 Conversion steps in the nitrification process (as presented in Colliver, 2000).

Even though N_2O is not present as an intermediate in the main catabolic pathway of nitrification, AOB are known to produce N_2O. This has predominantly been associated with denitrification capacity of AOB. AOB contain the enzymes to reduce NO_2^--N and NO with N_2O as final product. Note that these enzymes are the same as in regular denitrifying bacteria, but that in AOB denitrification is not associated with growth.

11.2.2 Denitrification

Denitrification is performed by a metabolically very diverse group of micro-organisms, bacteria as well as archaea, which couple oxidation of organic or inorganic substrates to reduction of nitrate, nitrite, NO and N_2O. As N_2O is an intermediate in the denitrification process, incomplete denitrification can lead to

N_2O emission. Many denitrifying micro-organisms are facultative denitrifiers, which preferentially use oxygen as electron acceptor, due to the higher energy yield. The different steps involved in the denitrification are presented in Figure 11.3.

	$2^- + 4H^+ + 4e^-$	→	$2NO_2^- + 2H_2O$
	$2NO_2^- + 4H^+ + 2e^-$	→	$2NO + 2H_2O$
	$2NO + 2H^+ + 2e^-$	→	$N_2O + H_2O$
	$N_2O + 2H^+ + 2e^-$	→	$N_2 + H_2O$
Total	$\mathbf{2^- + 12H^+ + 10e^-}$	→	$\mathbf{N_2 + 6H_2O}$

Figure 11.3 Conversion steps in the denitrification process (as presented in Otte, 2000).

11.2.3 Chemical reactions

Possible chemical pathways leading to N_2O formation in WWTPs are the reaction between nitrite and hydroxylamine leading to NO and N_2O and nitrite reductions with organic or inorganic compounds (Van Cleemput, 1998). In the first reaction the intermediate hydroxylamine production by AOB is required, complicating the distinction between chemical and biological N_2O production (paragraph, Kampschreur, 2009).

11.3 PROCESS PARAMETERS INFLUENCING N_2O FORMATION

Nitrous oxide emission has been extensively studied for soil systems. Reports about the emission of N_2O from activated sludge were only reported since the early nineties. An overview of all research on the emission of N_2O from WWTPs is presented in Table 11.1. In the same table an overview is presented of the research performed at laboratory scale.

From Table 11.1 it can be observed that there is a large variation in N_2O emission among the investigated WWTPs. This variation can be understood from the fact that N_2O can be formed both during nitrification and denitrification, and that different process parameters influence the formation of N_2O during these processes (see Figure 11.4[1]). The scale and frequency of changes in process conditions depend on the type of WWTP, reactor geometry, mixing intensity and aeration mode (Kampschreur, 2009).

In laboratory studies the process conditions typically change more rapidly than in large WWTPs, and this explains the higher level of variation in N_2O emission at laboratory scale as can be seen from Table 11.1. In addition more N_2O is stripped from laboratory scale systems than from full scale systems because aeration is often used as a means of mixing in laboratory reactors as well as for provision of oxygen.

---

[1]The process parameters mentioned will be discussed from page 10 onwards.

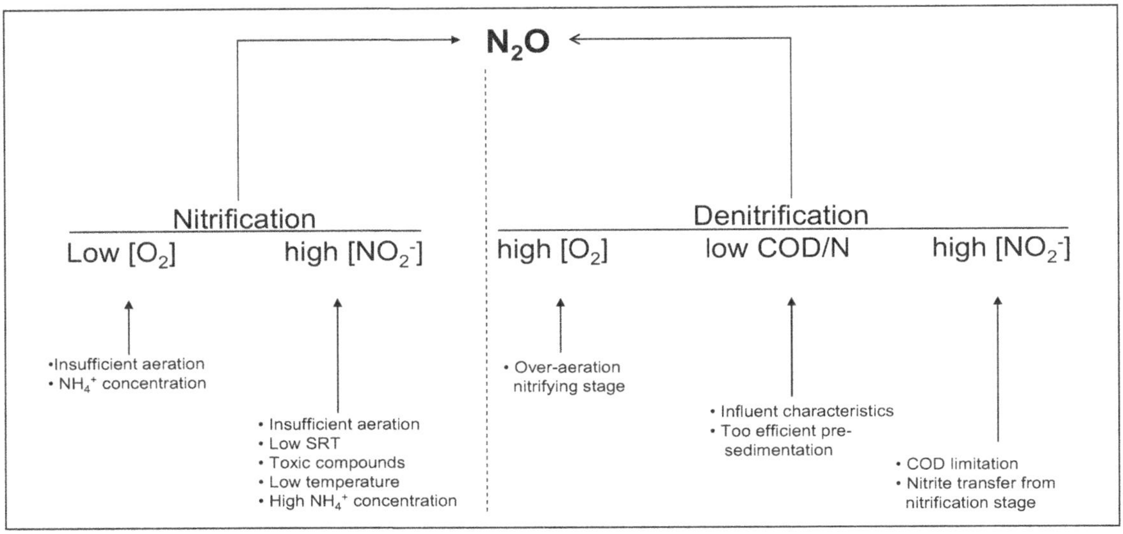

Figure 11.4 Main process parameters leading to N_2O formation and emission.

Oxygen concentration

A low dissolved oxygen concentration can lead to the formation of N_2O during the nitrification stage. Under these conditions autotrophic ammonia oxidizers use nitrite as the terminal electron acceptor to save oxygen for the oxygenation reaction of ammonia to hydroxylamine (Kampschreur, 2009). This process is referred to as the nitrifier denitrification. A low oxygen concentration can occur in the nitrification stage due to insufficient aeration or as a result of a high ammonium peak. The effect of a high ammonium peak was studied by Burgess et al. (2002) and Gjelsberg (1998), both found a decrease in the oxygen concentration followed by an increase in the nitrite concentration and N_2O emission.

Excessive aeration rates in the nitrification tank can lead to a substantial oxygen concentration in the denitrification tank. The consequence of this is that oxygen inhibits both synthesis and activity of denitrification enzymes. One of these, N_2O reductase is more sensitive to oxygen than the other enzymes, leading to N_2O emission during denitrification when oxygen is present in low amounts (Otte et al., 1996).

Nitrite

High nitrite concentrations in both nitrification and denitrification stages lead to an increase in N_2O emission. During nitrification increased nitrite concentrations lead to increased denitrification (effectively nitrite reduction to N_2O) by AOB provided ammonium is present (Colliver and Stephenson, 2000, Kampschreur, 2009). High nitrite concentrations during denitrification leads to a lower denitrification rate and accumulation of NO and N_2O (Schulthess et al., 1995). The nitrite concentration is affected by numerous operational parameters. These parameters are (Kampschreur, 2009):

- short SRT;
- toxic compounds (like sulphide)
- low(/high) temperatures
- high salinity
- sudden increase in ammonium load

Table 11.1 Overview of N_2O emission at full scale and laboratory scale (from Kampschreur, 2009).

Type of WWTP	N_2O emission (% of N-load)	Sample frequency	Remarks	Reference
		Full scale		
Activated sludge (11.000 p.e.)	0.035%[1]	Weekly grab sample (vented chamber) over 15 weeks		Czepiel et al. 1995
Twenty-five activated sludge plants	0–14.6% (0.6% average)	Single grab samples per WWTP	N_2O emission increased with increasing nitrogen load	Wicht and Beier, 1995
Activated sludge plant (60.000 p.e.)	0.001%	Twice-weekly grab samples over a year	N_2O emission increased with nitrite+nitrate concentrations	Sümer, 1995
Activated sludge plant (60.000 p.e.)	0.02%	Once or twice-weekly grab samples during 1.5 years		Sommer et al. 1998
Activated sludge plant (1.000 p.e.)	0.0–0.08%	On-line measurements during four aeration cycles (2 h)	N_2O emission decreased with proportionally shorter aeration periods	Kimochi et al. 1998
Nitritation – anammox sludge water treatment	2.3%[2]	On-line measurements during four days	N_2O emission increased with decrease in oxygen concentration (aerated stage) and increase of nitrite concentration (anoxic stage)	Kampschreur et al. 2008b
Nitrification stage (nitrogen removal stage) of activated sludge plant (620.000 p.e.)	4%	Three grab samples during one day		Kampschreur et al. 2008b
		Laboratory scale		
Continuous and batch tests denitrifying activated sludge – artificial wastewater	0–8%	Grab samples	N_2O emission increased with decrease of COD/N and increasing nitrite concentration	Hanaki et al. 1992
Continuous nitrifying activated sludge – artificial wastewater	2.3–16%	Daily grab samples	N_2O emission increased with decreasing oxygen concentration and SRT	Zheng et al. 1994
Batch test (10 h) denitrifying activated sludge – real wastewater	0–6%	On-line measurement	N_2O emission increased with increasing oxygen concentration	Schulthess et al. 1994

(Continued)

Table 11.2 Overview of N_2O emission at full scale and laboratory scale (from Kampschreur, 2009) (*Continued*).

Type of WWTP	N_2O emission (% of N-load)	Sample frequency	Remarks	Reference
Continuous oxic-anoxic SBR activated sludge – real high-strength wastewater (>50 d)	1-35%	Grab samples	N_2O emission higher with longer aeration period in one SBR cycle, probably linked to increased nitrite levels	Osada et al. 1995
Batch tests denitrifying activated sludge – real wastewater	1.2%	Grab samples	N_2O emission increased with decrease of pH	Thoern and Soerensson, 1996
Continuous oxic-anoxic SBR activated sludge – artificial wastewater (380 days)	5-95%	Grab samples	N_2O emission decreased over time – increased N_2O emission possibly a start-up phenomenon	van Benthum et al. 1998
Batch tests denitrifying activated sludge – artificial wastewater	0.005 -0.5%	Grab samples	N_2O emission increased with decrease of COD/N	Chung and Chung, 2000
Continuous nitrifying and denitrifying activated sludge – real wastewater	0.2-4.5%	Grab samples	N_2O emission decreased upon methanol addition for higher COD/N ratio	Park et al. 2000
Continuous oxic-anoxic SBR activated sludge – artificial high- strength wastewater	0.5 - >20%	On-line measurement	N_2O emission increased with decrease of COD/N	Itokawa et al. 2001
Continuous nitrifying activated sludge – artificial wastewater	0.08–1.17%	On-line measurement	N_2O emission increased with increasing ammonia shock loads and nitrite concentration	Burgess et al. 2002
Continuous activated sludge – real wastewater (50 days)	0.2–1.5%	Weekly grab samples	N_2O emission increased with decreasing oxygen concentration and decreasing SRT	Noda et al. 2003
Continuous anaerobic-anoxic SBR activated sludge – artificial wastewater	90%	On-line measurement	N_2O emission increased with increasing nitrite concentration – N_2O might be major product of denitrification by GAO[3]	Zeng et al. 2003
Continuous oxic-anoxic SBR activated sludge – artificial wastewater	>50%	On-line measurement	High N_2O emission (during oxic stage) was hypothesized to be caused by low diversity of population due to single carbon source	Lemaire et al. 2006
Continuous nitrifying activated sludge – artificial	0.2–0.5%	Daily grab samples	N_2O emission ratio increased with increasing salt concentration	Tsuneda et al. 2005

wastewater (30-300 days)				
Continuous denitrifying activated sludge – artificial wastewater (30-300 days)	0.005 – 0.02%	Daily grab samples	N_2O emission independent of salt concentration	Tsuneda et al. 2005
Continuous oxic-anoxic activated sludge – artificial wastewater (30-300 days)	0.7 -13%	Daily grab samples	N_2O emission increased with increasing salt concentration	Tsuneda et al. 2005
Batch test (5 h) nitrifying activated sludge- real wastewater	0.1–0.4%[4]	On-line measurement	N_2O emission is largest at 1.0 mg O_2/L and lower above and below this oxygen concentration – emission increases with nitrite concentration	Tallec et al. 2006a
Batch test (1 day) nitrifying biofilter - real wastewater	0.2–1% (0.4% average)	On-line measurement	N_2O emission ratio increased with decreasing oxygen concentration [5]	Tallec et al. 2006b
Batch test (1 day) denitrifying biofilter - real wastewater + methanol	% -1.3% (0.2% average)	On-line measurement	N_2O emission ratio increased with decreasing methanol addition	Tallec et al. 2006b
Batch test (5 h) nitrifying activated sludge- real wastewater	0.4%	On-line measurement	N_2O emission is largest at 0.3 mg O_2/L and lower above and below this oxygen concentration	Tallec et al. 2008
Continuous nitrifying SBR activated sludge – artificial wastewater	2.8%	On-line measurement	N_2O emission increased with decreasing oxygen and increasing nitrite concentration	Kampschreur et al. 2008a

[1]Based on the assumption of 100 g protein/person/day and 0.16 gN/gprotein (FAO-statistics, see also IPCC (2006)), the value of 3.2 gN_2O/person/year is converted into a ratio. This study is the basis for IPCC calculations.

[2]Sum of load based N_2O emission of nitritation reactor (1.7%) and anammox reactor (0.6%).

[3]GAO Glycogen Accumulating Organisms.

[4]Ratio is expressed per N-ammonia converted instead of N-load, because nitrate is present in the wastewater which is not converted here.

[5]The overall N_2O emission decreased with decreasing oxygen concentration because of the lower ammonia oxidation rate due to oxygen limitation.

COD/N

The COD/N ratio only affects the formation of N_2O during denitrification. A limited availability of biodegradable organic carbon is reported to increase the emission of N_2O (Kampschreur, 2009). One study showed the impact of various COD/N ratios (1.5; 2.5; 3.5 and 4.5); at the lowest COD/N ratio up to 10% of the nitrogen load was emitted as N_2O (Hanaki, 1992).

Rapidly changing process conditions

High concentrations of some of the above process parameters such as O_2 or NO_2^--N can be the result of rapidly changing process conditions. These changed process conditions can be caused by changes in environmental conditions or at the transition between anoxic and aerobic zones. Under these conditions it might be possible that the metabolism of bacteria needs time to respond to the changes in environmental conditions, resulting in substantial peak emissions of N_2O. Transient responses of activated sludge upon substrate availability were shown to be in the range of minutes (Vanrolleghem et al., 2004).

Bacterial populations subjected to continuously changing conditions can reduce their N_2O emission by adaptation; Alcaligenes *faecalis* was observed to reduce N_2O emission from 86% to 28% of nitrite converted after ten cycles of dynamic conditions (Schalk-Otte et al., 2000). Similar adaptation behaviour by mixed cultures was observed for N_2O emission upon exposure to toxic concentrations of formaldehyde (Garrido et al., 1998) and for N_2O emission during start-up of denitrification in a biofilm airlift reactor (van Benthum et al., 1998; Kampschreur, 2009).

11.4 EMISSION OF N_2O

From the previous paragraphs it becomes clear that N_2O can be formed during the nitrification or denitrification stage. This however does not mean that N_2O is emitted directly. This depends on the location where N_2O is formed and the solubility of N_2O. Nitrous oxide formed in anoxic zones can not be stripped there; on the other hand N_2O that is formed in aerobic zones (and not stripped immediately) can be converted in anoxic zones. The solubility of N_2O is relatively high (Henry coefficient is 24 mM/atm, while the Henry coefficient for oxygen is 1.3 mM/atm (Dean, 1992), and stripping is therefore not very fast. Dissolved N_2O in effluent can therefore lead to emission from receiving rivers and estuaries. Sommer et al. (1998) stated that during a winter period the N_2O dissolved in the effluent leaving a specific WWTP was five times higher than the N_2O emitted by air stripping due to the high solubility.

11.5 LOCATIONS CH_4 EMISSION AT WWTP

In contradiction to the formation processes of N_2O, the formation processes of CH_4 are very well known. However little information is available on the amount of CH_4 that is emitted from a domestic WWTP. Up till now one study performed by Czepiel et al. (1993) measured the emission of CH_4 from WWTPs.

Methane can only be formed under anaerobic conditions. In this respect emission of CH_4 can be expected from:

- sewerage systems
- influent works (formation in sewerage system, but emitted here);
- anaerobic / anoxic tanks as part of activated sludge systems;
- sludge digestion and handling;

Methane that is emitted from the influent works is formed in the sewer system. Little information is known about the amount of CH_4 formed in sewers. This is in contrast with H_2S. In most cases the influent headworks is totally covered and emission of CH_4 occurs after the air treatment. Due to the anaerobic conditions in anaerobic and anoxic tanks CH_4 might be formed. However, in the presence of aerobic zones methanogens will not survive.

At WWTPs that are equipped with an anaerobic sludge digester, CH_4 can be emitted from different locations related to the digester. Methane can be emitted during the different processes of biogas combustion (leakages, incomplete combustion) or during storage of digested sludge. At WWTPs without sludge digesters CH_4 formation is possible if the excess sludge is stored.

The amount of CH_4 emitted from these storage facilities will depend on the sludge retention time applied in the activated sludge system, the temperature and the level of dissolved CH_4, which in turn depends on the type of transport system prior to the WWTP.

Despite the presence of oxygen in the aeration tanks CH_4 can be emitted from these tanks. This is most likely CH_4 that has been formed earlier in the sewer or originates from rejection water from sludge handling facilities. The presence of methanogens in activated sludge has been proven by different authors (Lens et al., 1995, Gray et al., 2002). In these studies it was proven that the contribution of the CH_4 production by methanogens was very limited. Gray et al. (2002) reported 0.01–0.02% of the amount of carbon removed.

11.6 EMISSION FACTORS

The United Nations Framework Convention on Climate Change (UNFCCC) is the globally recognised basis for collective action on the reduction of anthropogenic greenhouse gas emissions (UNFCCC, 2007). One of the key obligations for signatory countries under the UNFCCC is the compilation of an annual national greenhouse gas (GHG) inventory, covering four general sectors (energy; industrial processes; agriculture, forestry and other land use; and waste). Emissions of CH_4 and N_2O from wastewater treatment and discharge are reported under the waste sector (IPCC, 2006b). However, GHG emissions are not usually measured directly, but instead are estimated through the application of formulas that link emissions to data on generally reported parameters (Foley and Lant, 2009).

11.6.1 Nitrous oxide

The revised 1996 IPCC Guidelines for National Greenhouse Gas Inventories (IPCC, 1997) estimation methodology for N_2O emissions from wastewater handling assumed minimal nitrogen removal occurs during treatment, and hence all influent nitrogen is discharged into rivers and/or estuaries, where it is mineralised, nitrified and denitrified under natural environmental processes. During these transformations, some of the discharged nitrogen will be emitted to the atmosphere as N_2O, at a default factor of 0.01 $kgN_2O\text{-}N/kgN_{discharged}$ (uncertainty range: 0.002–0.12) (IPCC, 1997). The 2006 IPCC Guidelines for National Greenhouse Inventories subsequently revised this default emission factor to 0.005 kgN_2O-N/kgN discharged (uncertainty range: 0.0005–0.25). The assumption that minimal nitrogen removal occurs in wastewater treatment plants (WWTPs) is incorrect for many countries. Recognising this, the 2006 IPCC Guidelines also updated the N_2O estimation methodology to include direct emissions from WWTPs with 'controlled nitrification and denitrification steps' (IPCC, 2006a). The proposed default emission factor was 0.0032 $kgN_2O\cdot person^{-1}\cdot yr^{-1}$ (uncertainty range: of 0.002–0.008), based on one full-scale study by Czepiel et al (1995) on a basic secondary treatment plant (without nitrogen removal) in New Hampshire, USA. Assuming a wastewater nitrogen loading of 16 $g\cdot person^{-1}\cdot d^{-1}$ for developed countries (i.e. high protein intake) (Tchobanoglous et al., 2003; IPCC, 2006a; DCC, 2008b), this equates to approximately 0.00035 kg N_2O-N/kg N. (Paragraph from Foley and Lant, 2009).

The above mentioned IPCC guidelines are used by most countries to estimate the emission of N_2O from domestic WWTP for their national inventory reports (NIR). Andrews et al. (2009) examined ten country-specific NIRs, six countries used the IPCC default procedures. The other four can be summarised as follows (Andrews et al., 2009):

- **Denmark** uses a country specific emission factor of 10.8 $gN_2O \cdot capita^{-1} \cdot y^{-1}$, with a quoted uncertainty of ±30%, which is just under 0.0024 kgN_2O-N/kg N load on secondary treatment, assuming the contribution to load on secondary treatment is 8 $gN \cdot person^{-1} \cdot day^{-1}$.
- **Japan** uses a country specific emission factor of 160 $mgN_2O\text{-}N/m^3$ flow to secondary treatment plus 0.6 $mgN_2O\text{-}N/m^3$ from the sludge which, using reasonable assumptions is equivalent to an emission factor of 0.004 kgN_2O-N/kg N load on secondary treatment, with an uncertainty of 146%.
- The **USA** uses the Czepiel et al. (1995) value of 3.2 $gN_2O \cdot capita^{-1} \cdot y^{-1}$ (0.0007 kg N_2O-N/kg N load on secondary treatment) for plants that only nitrify and 7 g $N_2O \cdot capita^{-1} \cdot y^{-1}$ (0.0015 kgN_2O-N/kg N load on secondary treatment) for plants that also denitrify.
- The **UK** (UKWIR Carbon Accounting Workbook (CAW)) uses a country specific emission factor of 0.002 kgN_2O-N emitted/kg N in sewage works load to estimate N_2O emissions from wastewater treatment. It has a high level of uncertainty, in the range of 30–300%.

11.6.2 Methane (CH_4)

The IPCC (IPCC, 2006a) prescribes the following generalised approach for estimating CH_4 emissions from wastewater treatment systems:

$$CH_4\ \text{Emissions}_w\ (\text{kg}) = \Delta COD_w \times EF \times MCF_w - R_{CH_4} \tag{1}$$

Similary, for CH4 emissions from sludge treatment systems:

$$CH_4\ \text{Emissions}_{sl}\ (\text{kg}) = \Delta COD_{sl} \times EF \times MCF_{sl} - R_{CH_4} \tag{2}$$

where

ΔCOD_w, ΔCOD_{sl}	=	Mass of chemical oxygen demand (COD) consumed / removed over the wastewater and sludge treatment processes, respectively (kg), determined by a simplified COD mass balance.
EF	=	Maximum CH_4 production / emission factor (0.25 kg CH_4 per kg COD).
MCF_w, MCF_{sl}	=	CH_4 correction factor for the type of process employed for wastewater treatment and sludge treatment, respectively (see Table 11.3)
R_{CH4}	=	Mass of CH_4 captured for combustion and/or flaring on the plant, or transfer out of the plant (kg).

In essence, this approach is a reconciliation of the estimated mass of CH_4 produced in the treatment process (i.e. ΔCOD × EF × MCF), with the measured mass of CH_4 captured in the associated biogas system (i.e. R_{CH4}). Any difference in these figures is assumed to be a loss of CH4 to the atmosphere. For treatment systems that are uncovered, R_{CH4} is zero. Notwithstanding the practical imprecision of measuring COD, biogas flow rates and biogas composition, it is clear that the accuracy of this estimation methodology is dependent upon the factor MCF.

Table 11.3 Default CH_4 correction factors for domestic wastewater (IPCC, 2006a; after Table 6.3).

Type of treatment	Comments	MCF	Range	CH_4 production (*EF x MCF*)
Centralised aerobic	Well managed	0.0	0.0–0.1	0.00 $kgCH_4 \cdot kg\ COD^{-1}$
treatment plant	Over-loaded	0.3	0.2–0.4	0.08 $kgCH_4 \cdot kg\ COD^{-1}$
Anaerobic digester or reactor	Does not include CH_4 recovery	0.8	0.8–1.0	0.20 $kgCH_4 \cdot kg\ COD^{-1}$
Shallow anaerobic lagoon	Depth < 2m	0.2	0.0–0.3	0.05 $kgCH_4 \cdot kg\ COD^{-1}$
Deep anaerobic lagoon	Depth > 2m	0.8	0.8–1.0	0.20 $kgCH_4 \cdot kg\ COD^{-1}$

(Information from Foley and Lant, 2009).

The above mentioned IPCC guidelines are used by most countries to estimate the emission of CH_4 from domestic WWTP for their national inventory reports (NIR). However, these guidelines neglect the emission of CH_4 that has been formed in the sewer and is stripped at the WWTP as the IPCC states that '*wastewater in closed underground sewers is not believed to be a significant source of methane'*. (IPCC, 2006a,b).

Andrews et al. (2009) examined 12 country-specific NIRs, eight countries used the IPCC default procedures (the 2006 Guidelines have not been applied yet for the preparation of NIRs) or IPCC procedures with the development of country-specific EFs. Sweden indicated that all secondary treatment was aerobic with no CH_4 emissions, and all sludge treatment had energy recovery and was dealt with under the energy sector. The UK methodology amounts to 20.8 kg CH_4/tonne raw DS for emissions from sewage treatment and onsite sludge treatment including digestion. Making reasonable assumptions this amounts to 0.6 kg $CH_4 \cdot capita^{-1} \cdot year^{-1}$. Canada uses an EF, based on IPCC (1997) of 4.015 $kg \cdot capita^{-1} \cdot year^{-1}$. Japan uses country specific values of 528.7 $mgCH_4/m^3$ of influent from wastewater treatment and 348.0 $mgCH_4/m^3$ from sludge treatment. This amounts to approximately 0.056 $kgCH_4 \cdot capita^{-1} \cdot year^{-1}$. An overview of all applied emission factors is presented in Appendix 2.1.

Chapter 12

Local regulation around GHG emission from WWTP

12.1 AUSTRALIA

12.1.1 Reporting Regulations

In Australia, the National Greenhouse and Energy Reporting System (NGERS) provides a single national framework for reporting greenhouse gas emissions, energy use and energy production by corporations. It is also intended to underpin a Federal Government emissions trading scheme at some stage in the future.

Under NGERS, any 'facility' using more than 100 terajoules (TJ) of energy, or emitting more than 25,000 tonnes of Scope 1 and Scope 2 carbon dioxide equivalents (CO_2-e) per annum is required to report to the Federal Department of Climate Change and Energy Efficiency (DCCEE)[1]. Similarly, any 'corporation' using more than 350 TJ of energy or emitting more than 87,500 tonnes CO_2-e in 2009/10 is required to report to the DCCEE. These corporate thresholds will decrease to 200 TJ and 50,000 tonnes CO_2-e for the 2010/11 reporting year. Changes in thresholds beyond 2011/12 are presently unknown.

The following Act and Regulations define the legislative requirements for greenhouse gas reporting within Australia:

- *The National Greenhouse and Energy Reporting Act 2007*, including amendments; and
- *The National Greenhouse and Energy Reporting Regulations 2008*, including amendments:
    - The Regulations provide detailed requirements for reporting under the Act, including definitions of operational control, facilities, the requirements for registration and the types of greenhouse gas emissions and energy consumption/production that have to be reported.

Further, the following documents provide methods and criteria for calculating greenhouse gas emissions and energy data under the *National Greenhouse and Energy Reporting Act 2007.*

[1] Under NGERS, reporting of Scope 3 emissions is voluntary.

- *The National Greenhouse and Energy Reporting (Measurement) Determination (2008):*
    - The *National Greenhouse and Energy Reporting (Measurement) Determination 2008* has been amended by the *National Greenhouse and Energy Reporting (Measurement) Amendment Determination 2009* (the Amendment Determination). The Amendment Determination is in general applicable to the 2009-10 reporting year; and
- *The National Greenhouse and Energy Reporting Technical Guidelines 2009* (or later revisions, if available).

These documents (and associated links) are publicly available via the DCCEE website (and associated links): www.climatechange.gov.au

The details of the NGERS are presented in Appendix 2.5.

12.2 FRANCE

In France there are no specific regulations concerning GHG emissions for the water sector. Nevertheless, all industries/companies with more than 50 employees must perform a carbon footprint assessment (new law July 2010). In the case of wastewater treatment plants, the French EPA (ADEME) recommends taking direct emissions of CH_4 and N_2O into account.

12.3 UNITED STATES OF AMERICA

In the United States of America (USA), environmental regulations may be enacted at the federal, state or local level. States or local authorities may issue regulations that are more restrictive than the federal government; however, those regulations apply only to those respective regions. The following summary of greenhouse gas legislation and regulations includes enacted regulations that affect, or may affect, publicly-owned treatment works (POTW). The summary includes current federal regulation and regulations by one state, California. No other states or coalitions of states have regulations on greenhouse gas (GHG) emissions that directly affect publicly owned wastewater treatment facilities. Proposed federal and state regulations are not discussed here.

12.3.1 Regulations that (may) affect publicly-owned treatment works (POTW)

On April 2, 2007, the Supreme Court of the United States, ruling in *Massachusetts vs. EPA, 2007,* found that GHGs, including carbon dioxide, are air pollutants covered by the federal Clean Air Act (CAA). On December 7, 2009, the USEPA Administrator signed two distinct findings regarding GHGs under the CAA a) that six, key, well-mixed GHGs threaten the public health and welfare of current and future generations and b) that the combined emissions of these well-mixed GHGs from new motor vehicles contribute to greenhouse gas pollution that threatens public health and welfare.

A series of clarifying actions then followed that culminated in the clarification of terms such as 'regulated New Source Review pollutant' and 'subject to regulation' and when a regulatory requirement to control emissions of a pollutant actually 'takes effect.' The action also confirmed that the next regulation dealing with GHGs was the light duty vehicle rule (LDVR) whose requirements would not 'take effect' any earlier than January 2, 2011.

The USEPA determined that one effect of the LDVR would be significant impacts on state permitting authorities for both Prevention of Significant Deterioration (PSD) and Title V permitting programs if these programs were left at current criteria pollutant trigger thresholds. PSD is that portion of the federal New Source Review program that regulates pollutants in regions of the country that have already 'attained' the national ambient air quality standard (NAAQS). PSD permitting was impacted since no national ambient air quality standard (NAAQS) for GHGs existed or will exist.

Title V was impacted because PSD is implemented through the Title V operating permit program, and GHGs can, by themselves, cause a facility to be a major source subject to Title V.

If the traditional 100/250 tons (USA) per year criteria pollutant trigger levels remained in place on January 2, 2011, they would lead to dramatic increases in the number of required permits – tens of thousands of PSD permits and millions of Title V permits – overwhelming permitting authorities unless the USEPA acted to 'tailor' the GHG thresholds upwards.

On May 13, 2010, the USEPA issued a final rule that 'tailors' GHG emissions from stationary sources under the CAA permitting programs and defines when permits under the PSD and Title V Operating Permit programs are required to address GHGs for new and existing industrial facilities. Facilities responsible for nearly 70 percent of the national GHG emissions from stationary sources will be subject to permitting requirements under this rule. The rule establishes a schedule that will initially focus CAA permitting programs on the largest sources that would have had to get PSD and Title V permits in any event, for their criteria pollutant emissions ('anyway sources'). The rule then expands to cover the largest sources of GHG emissions, at specific trigger levels, not previously covered by the CAA. Finally, it describes USEPA plans for additional, later steps in this process.

Compliance with PSD and Title V would be required by facilities that directly emit $\geq 75{,}000$ or $\geq 100{,}000$ tons of carbon dioxide equivalent (CO_2e) GHG emissions, depending on whether the emissions are from a new facility or from a change within an existing facility. In 2011 USEPA published a final rule, that for three years exempts biogenic CO_2 emissions from the above thresholds for PSD and Title V permitting purposes, while USEPA studies the issue further. Only biogenic CO_2 and not N_2O or CH_4 are covered by the temporary exemption.

12.3.2 Federal mandatory reporting of Greenhouse gases rule

In 2009, USEPA issued a rule that requires mandatory reporting of GHG emissions from large sources in the United States. The rule requires the collection and submission of accurate and comprehensive emissions data to inform future policy decisions. Suppliers of fossil fuels or industrial greenhouse gases, manufacturers of vehicles and engines, and facilities that emit 25,000 metric tons or more per year of GHG emissions are required to submit annual reports to USEPA.

The key reporting requirement applies to stationary combustion sources above 25,000 Million Tons (USA) CO_2e/yr. Stationary fuel combustion sources include, but are not limited to, boilers, simple and combined-cycle combustion turbines, engines, incinerators, and process heaters. The only wastewater treatment process emissions to be reported in this rule are those from onsite wastewater treatment located at industrial facilities. POTW process emissions have not been included in the rule because USEPA does not expect emissions from POTWs to exceed the applicable thresholds. The calculation of total emissions for the purposes of determining whether a facility exceeds the threshold does not include biogenic CO_2 emissions (e.g., those resulting from combustion of biofuels, including the elimination of biosolids from the calculation of combustion emissions).

However, to the extent POTWs utilize fossil fuel in their combustion facilities; these would be included in the threshold applicability determination. If the 25,000 MT (USA) CO_2e/yr threshold is exceeded based upon fossil emissions, then biogenic emissions would also need to be reported. There is no federal reporting requirement for combustion emissions below 25,000 MT CO_2,e/yr for a facility.

Emissions from stationary combustion units are determined using equations in the final rule, with the quantity of fuel combusted and emissions factors as the equation parameters. USEPA has clarified that utilities do not need to calculate or report emissions from combustion of biosolids. One key question for wastewater utilities is whether or not they have a facility whose collective combustion unit heat input rated capacity exceeds 30 mmBtu/hr. If this is so, they are required to calculate emissions for that facility to determine if they meet the threshold for reporting. For biogas combustion, only the CH_4 and N_2O portion of the emissions need be included, and for fossil fuels, all of the six well-mixed GHG constituents for which there are emission factors, need be included.

12.3.3 California's AB 32 the Global Warming Solutions Act of 2006

In 2006, the California Legislature enacted AB 32, the Global Warming Solutions Act of 2006. This Act establishes the first comprehensive GHG regulatory program in the USA, and commits California to achieving significant GHG emissions reductions by 2020. The Act will regulate all public and private entities that emit GHGs (defined as carbon dioxide, CH_4, N_2O, hydrofluorocarbons, perfluorocarbons, and sulfur hexafluoride) to achieve a state-wide emissions limit equivalent to the GHG emissions level back in 1990.This legislation directs the California Air Resources Board (CARB) to develop and implement a comprehensive program to achieve this goal by 2020. The reduction measures to meet the 2020 target are required to be largely adopted by the start of 2011. The CARB is required to coordinate with other state agencies to develop the AB 32 program. (The CARB is also charged with achieving a longer-range goal of complying with an earlier Executive Order of the Governor that sets state GHG emission target of 80% below 1990 levels by 2050.)

There are a number of elements required by AB 32, including reporting under a California Mandatory Reporting Program, and reductions prescribed in a comprehensive Scoping Plan. By January 1, 2008, CARB required annual reporting and verification of GHG emissions, beginning with the sources or categories of sources that contribute most to statewide emissions. The CARB's Mandatory Reporting Program uses reporting thresholds similar to the federal program, however, biogenic emissions do count in determining if the reporting threshold is exceeded, although they are reported separately from fossil emissions. Therefore, more POTWs will report in California compared to the federal program.

In the Scoping Plan, CARB has five regulatory measures directed at the water/wastewater sector including increasing water use efficiency, increasing water recycling, reduction in the magnitude and intensity of California's water systems, increased usage of urban runoff, increased renewable energy production from water systems and a public goods charge on water meters to pay for most of the above activities.

12.4 THE NETHERLANDS

Treatment of wastewater in the Netherlands is delegated to the Waterboard Authorities. These Waterboard Authorities are obliged to report the emission of greenhouse gases from wastewater treatment plants with a capacity larger than 136,360 p.e[2], or from wastewater treatment plants that treat more than 50 tonnes wet sludge (primary and secondary) per day (IPCC) from other WWTPs. The emission of N_2O and CH_4 are in these cases based on different emission factors as used in national inventory reports (NIR[3]). For N_2O an emission factor of 0.07% of the total nitrogen load is used. For CH_4 it is assumed that the majority of the CH_4 is emitted from sites that store digested sludge. In this case an emission factor of 18 g $CH_4 \cdot p.e.^{-1} \cdot year^{-1}$ is used.

---

[2]Based on 136 g total oxygen demand.

[3]Required under the Kyoto protocol. Emission factors used in the Netherlands are for N_2O 1 % of NKj influent and 0.007 kg CH_4 / kg $COD_{influent}$ for WWTP without sludge digestion and 0.0085 kg CH_4 / kg $COD_{influent}$ for WWTP with sludge digestion.

Since April 2010 the Dutch Waterboards signed a 'Dutch Climate Agreement' with the government. Part of this agreement is that the waterboards committed themselves to reduce the emission of non CO_2 GHG with 30% (equal to approximately 200 ktonnes CO_2e from 1990 to 2020). This number can be adjusted based on the outcomes of current research (performed from September 2010 – September 2011) on the emission of N_2O and CH_4 from WWTPs.

Chapter 13
Background and objectives research

13.1 AUSTRALIA

13.1.1 Emission of N_2O

In May 2008, WSAA commissioned The University of Queensland (UQ) to undertake field-based research to improve the level of certainty in the estimation methodologies for direct CH_4 and N_2O emissions from wastewater systems.

One of the aims of the study was to improve the level of certainty in the estimation methodologies for direct N_2O emissions from wastewater systems, as calculated in the NGERS Technical Guidelines. The default emission factor for wastewater treatment under NGERS (NGERS Technical Guidelines, 2009) is equivalent to 0.01 kg N_2O per kg nitrogen denitrified.

Specifically, the study addressed a knowledge gap identified in the earlier WSAA Literature Review (Foley and Lant, 2008): There is a high level of uncertainty in the magnitude and variability of N_2O emissions from biological nutrient removal (BNR) processes, under different physical configurations and process conditions.

The purpose of the research element was to determine the N_2O generation and emission rates in full-scale treatment plants, of varying physical configuration and under different process conditions. The study did not address the uncertainty in emissions from wastewater discharges to different receiving environments.

13.1.2 Emission of CH_4

With a global warming potential 25 times that of carbon dioxide, CH_4 is a potent GHG. Significant CH_4 production and emission from wastewater collection systems was not revealed until very recently. Indeed, the lack of data has led the IPCC to conclude, in its 2006 Guidelines (IPCC, 2006a,b), that '*wastewater in closed underground sewers is not believed to be a significant source of methane*'. As such, CH_4 in sewers is currently not considered in the accounting of greenhouse gas emissions from wastewater systems. However, the plentiful supply of readily biodegradable carbon, the high (>20°C) temperatures in certain locations and the presence of anaerobic biofilms, particularly in fully surcharged rising mains, suggest that there is potential for the formation of CH_4 in the raw sewage, and this is presently unaccounted in national GHG inventories (Foley and Lant, 2008).

For this reason several investigations were performed in Australia on the emission of CH_4 from rising mains. This research included measurements in the gas and liquid phase in the field and at laboratory scale. Based on the measurements two types of models were developed, one of which will be presented in this report. Further research was performed to study the impact of trade waste on the emission of CH_4 from rising mains. Finally research was performed to limit the formation of CH_4 in rising mains.

The methodology of the measurements in the gas and liquid phase in the field and at laboratory scale is discussed in section 14.5. The methodology of the other research topics is presented together with the results in section 15.5.

13.2 UNITED STATES OF AMERICA

13.2.1 Background

The push to achieve greater nutrient removal from wastewater treatment plant effluents has resulted in the development of a wide range of innovative biological nitrogen removal (BNR) processes. However, BNR strategies could be a potential contributor to atmospheric N_2O and NO depending upon the reactor configurations and operating conditions. In the future, as BNR is implemented at wastewater treatment plants, the flux of these gases to the atmosphere could increase.

Based on recent field-scale measurements, engineered BNR facilities, while effective to varying degrees in reducing *aqueous* nitrogen pollution, could emit up to 7% of the influent nitrogen load as *gaseous* N_2O and NO (Kampschreur et al., 2008b).

The WERF research project represents one of the first attempts at characterizing nitrogenous GHG emissions from wastewater treatment plants, and at developing a methodology for collection of full scale plant data from a range of nutrient removal facilities in the United States. Building on previous work by the project team, this information will be integrated into a mechanistic activated sludge process model, which will be refined through this project by the addition of autotrophic pathways for N_2O and NO emission. The refined mechanistic model will allow the industry to codify the results of this research, and develop a tool that will aid in the prediction and therefore mitigation of N_2O, NO and NO_2^--N emissions from WWTPs utilizing a range of wastewater treatment processes. Ultimately, this would allow the wastewater sector to engineer strategies for wastewater treatment that minimize gaseous nitrogen oxide emissions.

Although, from a fundamental perspective, N_2O and NO are known intermediates in heterotrophic denitrification (Knowles, 1982, Zumft, 1997) and autotrophic nitrification and denitrification (Anderson and Levine, 1986, Anderson et al., 1993, Kester et al., 1997, Ritchie and Nicholas, 1972, Stuven et al., 1992), the net contribution of BNR processes to N_2O emissions from wastewater treatment has only recently been explicitly acknowledged (USEPA, 2009). Based on the latest USEPA report on sources and sinks of N_2O from wastewater treatment operations, denitrification in anoxic zones is implicated as the dominant source of N_2O from BNR activated sludge reactors (USEPA, 2009).

However, nitrification could also play a role in N_2O generation and emission from WWTPs, especially under cycling between anoxic and aerobic conditions (Kampschreur et al., 2008b), as is common in BNR reactors. Therefore, N_2O fluxes from aerobic zones of WWTPs also need to be included in N_2O emissions inventories.

At this point, while there is considerable debate and interest on the 'potential' of constituent BNR processes for N_2O generation and emission, there are few reports that systematically quantify such emissions from full-scale BNR operations (Czepiel et al., 1995, Kampschreur et al., 2008b, Kimochi et al.,

1998, Sommer et al., 1998, Sümer et al., 1995, Wicht and Beier, 1995). Of these, only one has been conducted in the USA and focuses on a single non-BNR WWTP in New Hampshire (Czepiel et al., 1995).

Other full-scale studies have been conducted in Europe and have employed different methods for measuring N_2O emissions including the use of grab samples (Czepiel et al., 1995, Kampschreur et al., 2008b, Sommer et al., 1998, Sümer et al., 1995, Wicht and Beier, 1995) or online measurements (Kampschreur et al., 2008b, Kimochi et al., 1998). Given the broad diversity of BNR and non-BNR configurations that exist in the USA, a more detailed N_2O emissions database of WWTPs was needed, specifically obtained using a consistent protocol. Despite recognition of the possible role of BNR processes in N_2O emission, a measured database of N_2O emissions from these processes at the national scale does not currently exist. The WERF project focused on the quantification of N_2O emissions at twelve wastewater treatment plants (WWTPs) across the United States using a newly developed, USEPA-reviewed protocol. A subsequent focus was to determine the mechanisms and triggers of these emissions.

The principal motivation of conducting such a detailed monitoring campaign is the limited data currently used to 'estimate' the N_2O emissions of WWTPs. The current method is based on emission factor values of 3.2 g N_2O/population equivalents/year from non-BNR operations and 7.0 g N_2O/population equivalent/year for BNR operations (Czepiel et al., 1995, USEPA, 2009). Both emission factors are based on a limited data set and may not be broadly representative.

13.2.2 Emission factors United States of America

From a regulatory and policy perspective, organizations such as the USEPA are now beginning to explicitly include the contribution of BNR processes such as denitrification on the overall N_2O emission inventory from WWTPs (USEPA, 2009). A common approach to estimating the N_2O inventory of wastewater treatment processes is by using a 'single' emission factor with a value of 3.2 g N_2O/population equivalent/year and 7.0 g N_2O/population equivalent/year for non-BNR and BNR processes (USEPA, 2009). In these calculations, one population equivalent is defined as 100 gallons of wastewater discharged per capita per day. Conceptually, given that the inputs to a wastewater treatment plant and correspondingly the activity of the activated sludge bacteria are highly variable (Grady et al., 1999), it can be expected that there would be some degree of diurnal variability in N_2O emissions. Furthermore, owing to the fact that activated sludge bacteria have varying activities in different zones of the bioreactors (Grady et al., 1999), a certain degree of spatial variability in N_2O emissions is also expected from anoxic, anaerobic and even aerobic zones.

However, such spatial and diurnal variability in N_2O emissions is not considered in the simple emission factor approach adopted by the USEPA and IPCC for estimating the N_2O inventory of BNR and non BNR processes. Furthermore, by approximating the N_2O emissions from wastewater treatment processes using single emission factors across the board, the ability of certain operating conditions to selectively promote or minimize N_2O emissions is not understood.

Furthermore, not all emission factors are equivalent, owing to different conventions for normalizing, including wastewater flow rate, wastewater influent total nitrogen load or nitrogen load removed.

13.2.3 Objectives

The overall goal of this study was to quantify the emission of N_2O from WWTPs across the nation operated under different process conditions and configurations. The specific objectives were to:

- Develop a database of N_2O emissions fluxes from different activated sludge process configurations using a standard protocol.
- Identify key factors that are correlated with N_2O emission from activated sludge
- Determine the spatial and temporal variability in N_2O emissions from WWTPs (both BNR and non BNR) across the nation and examine the validity of the single emission factor approach in estimating the inventory of N_2O from WWTPs.

13.3 THE NETHERLANDS

Background

In the Netherlands the Dutch Waterboards aim to operate their wastewater treatment plants in a sustainable way. This implies that they try to minimize their energy consumption and (non CO_2) greenhouse gas emission. However, until 2007 very limited data were available on the emission of N_2O and CH_4 from Dutch WWTPs. For this reason there was a need to investigate the emission of GHG from Dutch WWTPs.

Objective

The objective of this research was to determine the level of N_2O and CH_4 emission from Dutch WWTPs and to understand the accuracy of the existing emission factors. In order to reach this objective the following research questions were raised:

- What is the emission of N_2O and CH_4 from a WWTP representative for the Dutch situation?
- Where do the emissions of N_2O and CH_4 originate from?
- Are the current emission factors for N_2O and CH_4 useful to estimate the total GHG emission from a Dutch WWTP?
- Is additional research required to indentify reliable emission factors for N_2O and CH_4?

Chapter 14
Methodology

14.1 NITROUS OXIDE MEASUREMENTS AUSTRALIA

14.1.1 Field sampling sites

The fieldwork program was conducted at seven full-scale BNR wastewater treatment plants in Australia. These plants were chosen to provide a range of plant sizes, process configurations, effluent qualities and climatic conditions. Their basic features are listed in Table 14.1. The widest possible range of WWTP types (in Australia) was selected with the expectation that differences between plant design and process conditions might help elucidate the factors influencing N_2O production.

14.1.2 Sample collection and analysis

For each WWTP, it was intended to conduct four sampling rounds (2–4 h duration each, morning and afternoon on two consecutive days). Due to a combination of circumstances, this was not possible at all plants. However, 20 of the intended 28 sampling rounds were completed over a five month timeframe in the Australian winter/spring of 2008.

For each sample round, data was collected to enable the construction of chemical oxygen demand (COD), total nitrogen (TN) and N_2O-N mass balances over the entire WWTP. The sampling locations and types of data collected are illustrated in Figure 14.1. Field data collection consisted of a combination of:

- Wastewater grab samples;
- Measurement of process conditions, namely temperature, pH, dissolved oxygen and oxidation-reduction potential (ORP), using a portable water quality meter (TPS 90FLMV); and
- Dissolved N_2O concentration measurements using a Clark-type microsensor (N_2O 25 with 70 μm outside tip diameter, Unisense A/S, Aarhus, Denmark), logged via a Pico ammeter to a laptop.

Table 14.1 Wastewater Treatment Plant Sites.

Type of process	Sludge load[1]	Temperature	SRT (appr.)	Effluent TN[1]	Process description
	($kgTKN_{in} \cdot kg\ d.w.^{-1} \cdot d^{-1}$)	(°C)	(d)	(mg/l)	
1. Ox.Ditch (4 rounds)[2]	0.043 0.042 0.03 0.042	23	13	3	Inlet works, anaerobic contact tank, 2 × extended aeration oxidation ditches (in parallel) with diffused aeration, secondary sedimentation; mechanical sludge thickening and dewatering.
2. Johannesburg (3 rounds)	0.012 0.01 0.013	21	20	5	Inlet works, 2 × extended aeration Johannesburg bioreactors (in parallel) with submerged aspirating OKI™ aerators, secondary sedimentation, sludge thickening, aerobic digestion, mechanical and solar dewatering.
3. SBR (3 rounds)	0.021 0.017 0.021	20	16	15	Inlet works, primary sedimentation, sequencing batch reactor (4 compartments) with diffused aeration and bio-selector zone, sludge thickening, anaerobic digestion and mechanical dewatering.
4. MLE-1 (1 round)	0.031	21	13	12	Inlet works, primary sedimentation, 11 × covered Modified Ludzack Ettinger (MLE) bioreactors (in parallel) with diffused aeration, secondary sedimentation, sludge thickening, mechanical dewatering and lime stabilisation.
5. MLE-2 (4 rounds)	0.06 0.065 0.072 0.07	20	8	11	Inlet works, primary sedimentation, 2 × Integrated Fixed Film Activated Sludge bioreactors + 1 × MLE bioreactor (in parallel) with diffused aeration, secondary sedimentation, sludge thickening and anaerobic digestion.
6. MLE-3 (3 rounds)	0.01 0.01 0.012	18	15	13	Anaerobic lagoon, 1 × MLE bioreactor with diffused aeration, secondary sedimentation, maturation lagoons, and sludge wasting to an aerated facultative lagoon.
7. A^2/O (2 rounds)	0.009 0.01	21	14	3	Inlet works, 2 × parallel trains: 1) Primary sedimentation, 4-stage Bardenpho bioreactor with diffused aeration, secondary sedimentation; 2) Pre-fermenter, four-stage bioreactor (similar to A^2/O configuration) with diffused aeration and supplemental COD dosing by primary sludge from Train 1, secondary sedimentation, tertiary filtration, sludge thickening, aerobic digestion, sludge lagoon and mechanical dewatering.

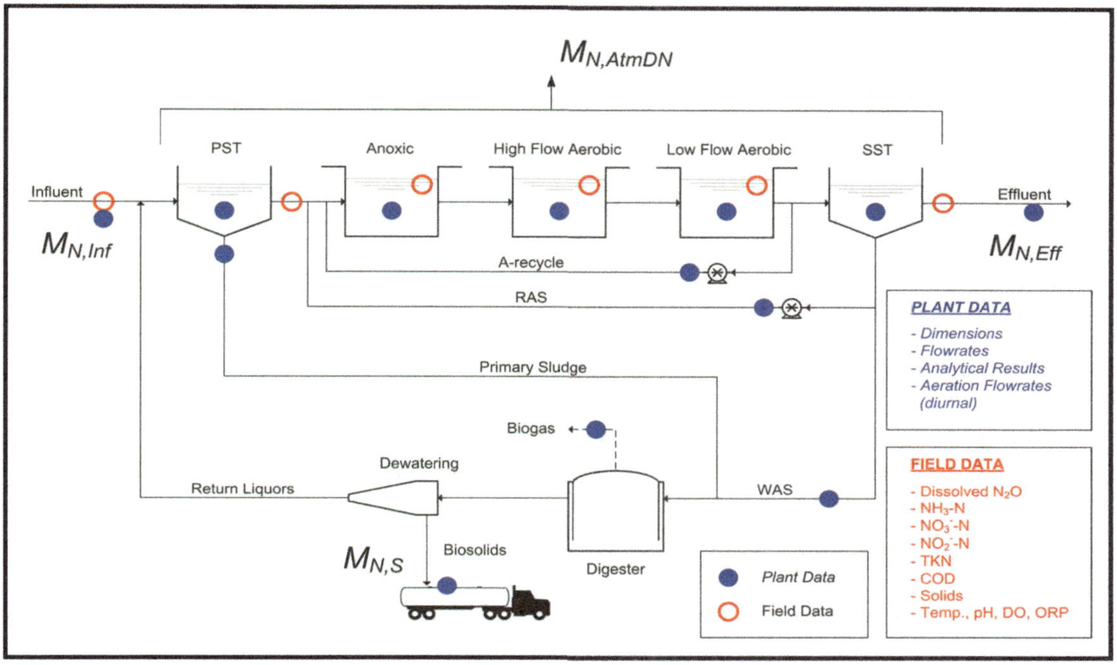

Figure 14.1 Typical data (types and locations) collected for the construction of the COD and total nitrogen mass balances over the entire process at each WWTP.

14.1.3 Determination of N_2O emissions

Chemical oxygen demand and Total Nitrogen Mass Balances over Entire WWTP Processes

At each of the seven WWTP sites, total COD and TN mass balances were constructed across the entire process. These mass balances drew upon the analytical data collected in the field (i.e. COD, TKN, $^{-}$-N, NO_2^--N, MLSS, MLVSS concentrations), as well as the plant data (i.e. flowrates, reactor volumes, solids capture efficiencies, biosolids tonnages and composition, biogas production and composition) supplied by the WWTP operators (refer to Figure 14.1). The purpose of this initial mass balance analysis was to:

- Ensure an accurate characterisation of the WWTP operation, such that both COD and TN balances over the WWTP generally achieved greater than 90% closure; and
- To determine the mass of nitrogen denitrified and emitted to the atmosphere, according to Equation 3:

$$M_{N,AtmDN} = M_{N,Inf} - M_{N,Eff} - M_{N,S} \tag{3}$$

Where:

$M_{N,AtmDN}$ = Mass of N denitrified to atmosphere, either as N_2 or N_2O gas ($kg.d^{-1}$)
$M_{N,Inf}$ = Mass of nitrogen in influent (kg/d)
$M_{N,Eff}$ = Mass of nitrogen in effluent (kg/d)
$M_{N,S}$ = Mass of nitrogen in wasted solids (kg/d)

Equation 3 assumes that the WWTP is operating at steady-state, with no net accumulation of nitrogen within the biomass inventory. All of the plants investigated had medium to long solids retention times (SRT) (i.e. 8–20 d). Hence, the change in biomass inventory is relatively slow, and the assumption of near steady-state conditions should hold for the two day sampling period at each WWTP.

Given the large variation in physical size and treated load of the seven WWTPs surveyed, Equation 3 provided a means of normalising the generation and emissions of N_2O (i.e. as a percentage of the total nitrogen denitrified to atmosphere).

Liquid Phase N_2O Mass Balances over Individual WWTP Zones

The second phase of mass balance analysis examined bulk liquid phase N_2O across the individual zones of each WWTP. For five of the seven sites, the WWTP was divided into five reactor zones for this mass balance analysis (i.e. primary sedimentation tank or anaerobic zone; anoxic zone; highly aerated aerobic zone; less aerated aerobic zone; and secondary sedimentation tank). At WWTP No.6, the plant was divided into seven reactor zones (three anoxic zones, three tapered flow aerobic zones, and secondary sedimentation tank). At WWTP No.3 (SBR), the mass balance was divided across each operational phase (i.e. fill/aerate, settle, decant). The general formulation of the mass balance construction is given in Equation 4:

$$\frac{dM_{N_2O-N,R}}{dt} = \sum M_{N_2O-N,In} - \sum M_{N_2O-N,Out} - Tr_{N_2O-N,R} + G_{N_2O-N,R} \tag{4}$$

Where:

$\frac{dM_{N_2O-N,R}}{dt}$ = change in mass of N_2O-N in the reactor zone, over time ($kg.d^{-1}$)

$\sum M_{N_2O-N,In}$ = sum of *i* mass flows of N_2O-N into the reactor zone ($kg.d^{-1}$)

$\sum M_{N_2O-N,Out}$ = sum of *j* mass flows of N_2O-N out of the reactor zone ($kg.d^{-1}$)

$Tr_{N_2O-N,R}$ = mass transfer of N_2O-N from the reactor liquid to gas phase ($kg.d^{-1}$)

$G_{N_2O-N,R}$ = net generation of N_2O-N in the reactor zone ($kg.d^{-1}$) (i.e. net result of N_2O-N production and consumption due to biological reactions in the reactor)

Similar to Equation 3, it was assumed that the reactor zones operate at near steady-state conditions, and are well-mixed. Equation 4 can then be expanded and re-formulated to solve for $G_{N_2O-N,R}$:

$$G_{N_2O-N,R} = \sum_j Q_{Out,j} \times [N_2O-N]_R - \sum_i Q_{In,i} \times [N_2O-N]_{In,i} + V_R \times k_L a \times \left([N_2O-N]_R - [N_2O-N]_S^*\right) \tag{5}$$

Where

$Q_{In,i}$, $Q_{Out,j}$ = individual flows in and out of the reactor zone (Ml/d)

$[N_2O\text{-}N]_{In,I}$ = concentration of N_2O-N in the incoming streams (mg/l or kg/Ml), which is generally equal to the N_2O-N concentration in the originating reactor

$[N_2O\text{-}N]_R$ = concentration of N_2O-N in the reactor zone (mg/l or kg/Ml)

V_R = volume of the reactor zone (Ml)

k_La = volumetric mass transfer coefficient (d^{-1})

$[N_2O\text{-}N]_s^*$ = saturation concentration of N_2O-N in water at atmospheric conditions = 2.57×10^{-4} kg/Ml at 20°C (Weiss and Price, 1980)

This calculation was completed for each reactor zone at each WWTP. It was then repeated at each reactor zone for the "best-case" combination of the lower limit values of the 95% confidence intervals of the measured N2O-N concentrations and estimated mass transfer coefficients (refer to Figure A.6.1). The calculation was then repeated again for the 'worst-case' combination of the upper limit values of the 95% confidence intervals. These calculations determined the uncertainty range of net N2O-N generation (i.e. production minus consumption) in each reactor zone. Negative values of $G_{N_2O-N,R}$ indicate N2O-N consumption is greater than N_2O-N production in that particular zone.

The net generation and emissions of N_2O-N in each reactor zone were then summed to give the net generation, $G_{N_2O-N,WWTP}$, and emissions, $Tr_{N_2O-N,WWTP}$, of N_2O-N for the whole WWTP:

$$G_{N_2O-N,WWTP} = \sum G_{N_2O-N,R} \tag{6A}$$

$$Tr_{N_2O-N,WWTP} = \sum Tr_{N_2O-N,R} = \sum\left\{V_R \times k_La \times \left([N_2O-N]_R - [N_2O-N]_S^*\right)\right\} \tag{6B}$$

The ratio of N_2O-N mass transfer emissions to net generation was then calculated for each WWTP:

$$\%Tr_{WWTP} = \frac{Tr_{N_2O-N,WWTP}}{G_{N_2O-N,WWTP}} \times 100\% \tag{7}$$

Normalisation of Nitrous Oxide Mass Balance Results

To compare results between different reactor zones, $G_{N_2O-N,R}$ was normalised by dividing by the corresponding total mass of nitrogen denitrified to the atmosphere from the entire WWTP:

$$GF_R = \frac{G_{N_2O-N,R}}{M_{N,AtmDN}} \tag{8A}$$

Where:

GF_R = N_2O-N generation factor for each reactor zone ($kgN_2O\text{-}N/kgN_{denitrified}$)

$M_{N,AtmDN}$ is calculated according to Equation 1.

Similarly, the mass transfer emissions to atmosphere from each reactor zone, $Tr_{N_2O-N,R}$, were normalised by dividing by $M_{N,AtmDN}$:

$$EF_R = \frac{Tr_{N_2O-N,R}}{M_{N,AtmDN}} \tag{8B}$$

Where:

EF_R = N_2O-N emissions factor for each reactor zone ($kgN_2O\text{-}N/kgN_{denitrified}$)

To compare results across sites, $G_{N_2O-N,WWTP}$ for each WWTP was normalised by dividing by its corresponding total mass of nitrogen denitrified to the atmosphere:

$$GF_{WWTP} = \frac{G_{N_2O-N,WWTP}}{M_{N,AtmDN}} \tag{9}$$

Where:

GF_{WWTP}= N_2O-N generation factor for entire WWTP ($kgN_2O\text{-}N/kgN_{denitrified}$).

The determination of the Nitrous Oxide Volumetric Mass Transfer Coefficients is presented in Appendix 2.6.

14.1.4 Quality control

The electrochemical microsensor was two-point calibrated before and after each field sampling round at ambient temperature, using distilled water (zero point) and a freshly prepared 0.15 mM N_2O solution. Laboratory trials with the N_2O25 microsensor in 0.27 mM N_2O solution showed its maximum measurement error to be $\pm$ 0.3%. The response of the electrochemical microsensor is known to be linear in the range of 0 - 1.2 mM (Andersen et al., 2001). At each sampling location, the microsensor was fully immersed in the reactor and allowed to stabilise and log data at 3 s intervals over a 5 - 10 minute period. The signal data from this sampling period at each location was then statistically analysed to calculate its 95% confidence interval (t-dist, $\alpha = 0.05$). The microsensor was also zero point calibrated with distilled water in the field after every measurement to provide a baseline signal and correct for any drift.

The TPS 90FLMV water quality meter was calibrated before and after the two-day sampling exercise at each WWTP. Grab samples for soluble species were immediately filtered using 0.22μm syringe filters, acid-preserved and kept on ice before analysis.

COD was measured by the colorimetric method described in APHA (1995) using commercial Lovibond tubes in a range of 0 to 150 $mgCOD.L^{-1}$. The ammonium (NH_4^+-N), nitrate ($^-$-N) and nitrite (NO_2^--N) concentrations were analysed using a Lachat QuikChem8000 Flow Injection Analyser (Lachat Instrument, Milwaukee, USA). Total Kjeldahl Nitrogen (TKN), mixed liquor suspended solids (MLSS) and volatile solids (MLVSS) were analysed according to Standard Methods (APHA, 1995).

Physical plant data (e.g. reactor dimensions, plant flowrates, diurnal aeration flowrates) were provided by the WWTP operators for the specific days and times of field sampling.

Where possible, the WWTP operators also supplied their own routine process and analytical data that provided a useful cross-check against results from the grab samples collected during the field study.

14.2 NITROUS OXIDE MEASUREMENTS FRANCE

14.2.1 Field sampling sites

Four WWTPs have been monitored, all of them with N removal: two 'classical' BNR activated sludge facilities, one 'plug -flow' process and one Membrane Bioreactor (MBR). The characteristics of these WWTPs are summarized in Table 14.2.

Table 14.2 Characteristics four WWTPs studied in the N_2O research.

WWTP	Water line	Biological treatment
WWTP 1	Pre treatment- Activated Sludge – Secondary Clarifier	Three-zones completely Stirred Tank Reactor. (anaerobic- anoxic- aerobic with sequenced aeration)
WWTP 2	Pre treatment - Activated Sludge - Secondary Clarifier	Plug flow process (anoxic –aerobic)
WWTP 3	Pre treatment - Activated Sludge - Secondary Clarifier	Three-zone completely Stirred Tank Reactor. (anaerobic- anoxic- aerobic with sequenced aeration)
WWTP 4	Pre treatment – Activated Sludge – Membrane Bio Reactor	Activated sludge with sequenced aeration followed by MBR

The configuration of WWTP 1–4 are presented in Appendix 2.7.

14.2.2 Sample collection and analysis

The way the samples were taken for N_2O analysis depended on if the surface was aerated or non-aerated. Both protocols are described below.

Sample collection from aerated surfaces

The square wooden sampling box (1m × 1m) is used in aerated conditions (see Figure 14.2). A pipe connected to a gas mass counter allows the flow of air supplied by the aeration system - 'real' Q - to be collected and quantified (in m^3/h). A spur from this pipe allows the sample of air to be directed to the gas analyzer.

Figure 14.2 Sampling box for aerated surfaces.

Sample collection from non-aerated surfaces

In case of non-aerated surfaces the sampling system used was the Odoflux dynamic flow chamber (see Figure 14.3).

Figure 14.3 Sampling box for non-aerated surfaces.

The flow chambers covered an area to be studied as hermetically as possible in order to isolate the surface from external conditions. The gases emitted by the isolated surface are collected by a vector gas which is injected into the chamber (Q1). The mixture of vector gas and gaseous effluent is then collected for analysis. The flow chamber used consists of an acrylic resin cylinder capped by a hemisphere, also of acrylic resin. The air-tightness of these two parts is ensured by polypropylene screws. The sweep air supply is fed by a pipe coiled against the wall and pierced with holes (flow Q1). The air sample is taken at the top of the hemisphere by means of a probe which has several holes along its length in order to ensure homogeneous sampling. The air is sucked in at flow Q2, of which 1l/min is directed to the analyser; the other part (Q2–1l/min) is released into the atmosphere.

To respect 'pseudo-isokinetic' conditions, the applied flow Q1 must be equal to the flow sucked in Q2. The protocol developed in the framework of this project for measuring GHG recommends a relatively weak flow (Q1 = Q2) of the order of 3 to 10l/min; this flow thus ensures 'static chamber' conditions in which the surface of the liquid is not disturbed and the surface emissions are very slightly diluted.

Analysis

The measured gas is sampled at the level of each spur using semi-supple teflon pipe, 4/6 mm in diameter. The measurement chain detailed below is illustrated in Figure 14.4.

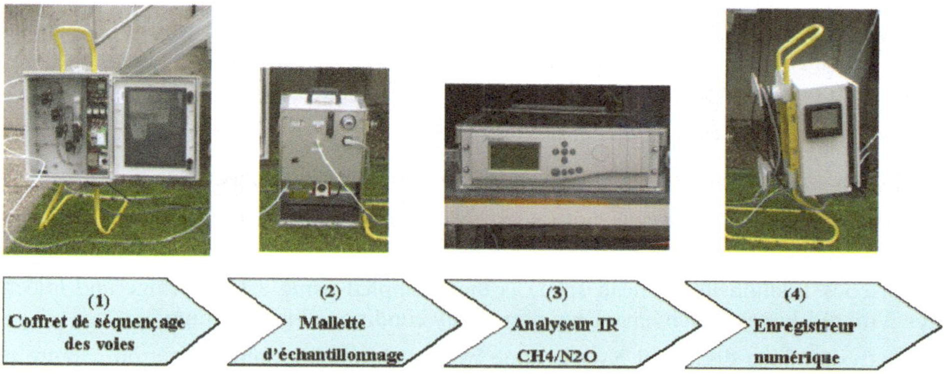

Figure 14.4 Measurement chain to determine N_2O concentration in gas samples; (*English version of the picture was not available, measurement chain is explained in the text).*

The collected gas is carried to the first module of the mobile analysis system, i.e. the measurement channel scrutiny system (1). On leaving this channel sequencing box, a single measurement channel is provided in the form of an exit via a semi-supple teflon pipe, 4/6 mm in diameter; this pipe is then connected to a sampling unit (2) which ensures the pre-conditioning of the measurement gas (Cooling of the gas using the Peltier effect, to reduce the dew point of the measurement gas and its exit regulation at a value of +5°C, Fine filter, with a 2µm filter cartridge, with humidity detection, flow-meter to ensure exact regulation of the flow of collected gas). When the collected gas leaves the portable sampling unit, it is then connected to the Servomex model 4210 gas analyser (3). The Servomex model 4210 analyser uses infra-red technology with gas filter correlation (GFC) for the compounds N2O and CH4. The data are taken continuously by the numerical recorder model QX/ HONEYWELL (5). The data thus recorded may be exported using Excel. The concentrations measured are expressed in ppm volume/volume.

Gas flow and N_2O emission

The resulting area flow of gas measured is calculated from the concentration measured at exit from the sampling system (N2O or CH4 in mg/m^3 air) and from the air flow applied to the sampling system (m^3/h).

The flow is related to the sampling area (area of the *flow chamber* in the case of non-aerated surfaces or the area of the *sampling box* in case of aerated surfaces), and is expressed in $mg{\cdot}h^{-1}{\cdot}m^{-2}$.

Additional data

In order to understand the formation of N2O during the treatment of waste water, the appearance of the different nitrous forms in biological treatment tanks, i.e. ammoniac (NH_4^+-N), nitrates ($^-$-N) and dissolved N_2O were monitored. These measurements would appear to be very necessary for monitoring the dynamics of the formation of N_2O emitted during waste water treatment processes. It likewise contains the protocol for measuring dissolved N_2O which was developed by the Rennes CEMAGREF and transferred to CIRSEE.

Ammonium and nitrate analysis

A VARIONPlus NH_4^+-N / $^-$-N (WTW) probe allows continuous measurement of these different nitrous forms in water. It uses specific ion measurements, which is a potentiometric method.

Dissolved N_2O

The protocol for the analysis of dissolved N_2O presented here was drawn up by the Rennes CEMAGREF. The water samples are taken and conditioned in 330 ml flasks, sealed hermetically with a septum. The sample conditioning stages are detailed below:

- Weigh 100 g of water sample in a 330 ml glass flask.
- Add 1 drop of sulphuric acid and close the flasks hermetically (stopper + septum).
- Place the samples in a container with warm water at 80 °C for 1 hr, the dissolved N_2O is then released into the head space.
- The gaseous fraction in the head space is then sampled using a gas syringe and injected into three 3 ml pill-boxes which have been previously conditioned in a vacuum;
- The analysis of the dissolved N_2O is done by gas chromatography - electron capture detector (GC/ECD) by the Rennes CEMAGREF, in France.

14.3 NITROUS OXIDE MEASUREMENTS USA

14.3.1 Field sampling sites

Nitrous oxide emissions were monitored during this study from a wide range of activated sludge processes (both non-BNR and BNR). The modes of operation are summarized in this section according to the process configuration. For the sake of confidentiality, the specific locations of these processes are not be described. However, for geographic representativeness, it was ensured that these processes were broadly distributed around the North-East (4), Mid-Atlantic (2), Mid-West (2) and South-West (4) regions of the United States. The dimensions of the process bioreactors sampled and the sampling locations are further provided in Appendix 2.2 of this report. A description of the type of sampled processes is presented in Table 14.3.

Table 14.3 Overview of type of processes sampled in N_2O research.

Plant Configuration	Description
Separate-stage BNR	The low-rate separate-stage nitrification denitrification process at this WWTP was sampled. The process was configured as a sequence of five reactors in series, as shown in Appendix 2.2. The influent to this process consisted of the clarified effluent from an upstream high-rate process, mainly engaged organic carbon removal. The influent was fed in a step-feed fashion to the first two aerobic zones. The last three zones of this process were non-aerated and the second non-aerated zone received methanol to promote denitrification. The effluent channel of this process was aerated prior to secondary clarification.
Four-stage Bardenpho	The four-stage Bardenpho process consisted of pre-denitrification (without external carbon addition) followed by a primary aerated zone, as shown in Appendix 2.2. The effluent of the primary aerated zone was internally recycled to the anoxic zone. Following the primary aerated zone was a de-oxygenation zone to scavenge dissolved oxygen, prior to methanol addition for enhanced denitrification. The final zone in this process was aerated primarily for stripping off the dinitrogen gas produced during denitrification, prior to secondary clarification.
Step-feed BNR 1	The four-pass step-feed BNR process sampled consisted of pre-anoxic zones comprising about 1/3 of the pass volume followed by aerated zones, as shown in Appendix 2.2. The transition zone between each pass was non-aerated to facilitate deoxygenation. The approximate influent flow split was10%-40%-30%-20% to

	passes A, B, C and D, respectively. Pass A also received pre-settled anaerobic digestion centrate, which constituted approximately 30% of the influent TKN load to the process. Return activated sludge was also fed to Pass A.
Step-feed non-BNR	The step-feed non-BNR process sampled was configured and operated in four-pass step-aeration mode. The process was completely covered primarily for odor control. The headspace off-gases were consolidated and fed to a biofilter. The approximate influent flow split was10%-40%-30%-20% to passes A, B, C and D, respectively. Return activated sludge was fed to Pass A.
Separate centrate	The separate centrate treatment process was operated to process pre-settled anaerobic digestion centrate and partially convert the influent NH_4^+-N to NO_2^--N. The separate centrate treatment process was operated in plug flow mode, as shown in Appendix 2.2. Effluent from the separate centrate tank was fed to the overall plant return activated sludge line for possible bioaugmentation with primarily ammonia oxidizing bacteria (AOB) and for nitrogen removal via the short-cut nitrite pathway.
Plug-flow 1	The first plug-flow process sampled was designed and operated primarily for organic carbon removal and nitrification and did not have dedicated anoxic zones or external organic carbon addition. The process was configured in four-pass mode, as shown in Appendix 2.2.
Plug-flow 2	The second plug-flow process sampled was also designed and operated for organic carbon removal and nitrification and did not have dedicated anoxic zones or external organic carbon addition. The process was configured in two-pass mode, as shown in Appendix 2.2.
MLE 1	The first modified Lutzack Ettinger (MLE) process sampled was originally designed for operation in enhanced biological phosphorous removal mode, but subsequently operated in MLE mode. As shown in Appendix 2.2, the process consisted of pre-denitrification without external organic carbon addition. Biomass from zone 6, which was aerated, was internally recycled to anoxic zone 2. Return activated sludge was fed to anoxic zone 1.
MLE 2	The second modified Lutzack Ettinger (MLE) process sampled was also originally designed for operation in enhanced biological phosphorous removal mode, but subsequently operated in MLE mode. As shown in Appendix 2.2, the process consisted of pre-denitrification without external organic carbon addition. Biomass from zone 6, which was aerated, was internally recycled to anoxic zone 2. Return activated sludge was fed to anoxic zone 1.
Step-feed BNR 2	The second step-feed process sampled was configured in four-pass mode as shown in Appendix 2.2. Each pass consisted of pre-anoxic zones comprising 1/3 of the pass volume followed by aerobic zones. The approximate influent flow split was50%-30%-20%-0% to passes A, B, C and D, respectively. The anoxic zones were mixed via low intensity pulse aeration. The return activated sludge was fed to Pass A.
Oxidation ditch	The oxidation ditch process shown in Appendix 2.2 was operated to achieve simultaneous nitrification and denitrification by operation are uniformly low aeration intensities and dissolved oxygen concentrations. The influent flow to the process was fed to the inner loop and was mixed and circulated using surface mixers. No external organic carbon was added to enhance denitrification. Return activated sludge was fed to the inner loop of the process.
Step-feed BNR 3	The third four-pass step-feed BNR process sampled consisted of pre-anoxic zones comprising about 1/3 of the pass volume followed by aerated zones, as shown in Appendix 2.2. The approximate influent flow split was 33.3%-33.3%-33.3%-0% to passes A, B, C and D, respectively. Pass A also received pre-settled anaerobic digestion centrate, which constituted approximately 40% of the influent TKN load to the process. Return activated sludge was also fed to Pass A. The reactors of this process were also covered and thus only composite measurements of the overall headspace could be performed.

14.3.2 Samples collection and analysis

One of the most significant outputs from the WERF project has been the development of a protocol that enables collection of N_2O emission fluxes from open-surface activated sludge bioreactors using consistent methodology. The protocol was submitted to the USEPA for review during late 2008 and the comments provided have been duly incorporated herein. The protocol is intended to provide utilities and field sampling teams with a detailed description of the data collection methodology and analysis requirements to enable calculation of gaseous nitrogen fluxes from different zones of activated sludge trains in a wastewater treatment facility.

The protocol was officially released during the 2nd Water Environment Federation Nutrient Removal Conference in Washington DC during June 2009 and is also available online at www.werf.org.

Sampling design for Full-Scale monitoring

The N_2O emission fluxes of several wastewater treatment plants that are accomplishing nitrification and denitrification were determined. Testing was conducted at each plant during which gas phase monitoring was performed in real-time continuous mode and liquid phase sampling was performed via discrete grab sampling. Plants were subjected to two campaigns conducted once in warm temperature conditions (i.e. summer, early fall), and cold temperature conditions (winter/early spring) in the Northeast and Midwest and twice in plants along the West Coast (Fall and Spring), not subject to significant temperature changes.

Sampling procedures-headspace gas measurement

The overall procedure for measuring N_2O, NO and NO_2^--N fluxes from the head-space of activated sludge tanks involves a variant of the EPA/600/8-86/008 and the South Coast Air Quality Management District (SCAQMD) tracer methods. This variant was developed to measure those sources that have a relatively high surface flux rate when compared to diffusion (for instance, spilled oil containment).

Commercially available replicas of the US EPA surface emission isolation flux chamber (SEIFC, Figure 14.5) were used to measure gaseous N fluxes from activated sludge reactors.

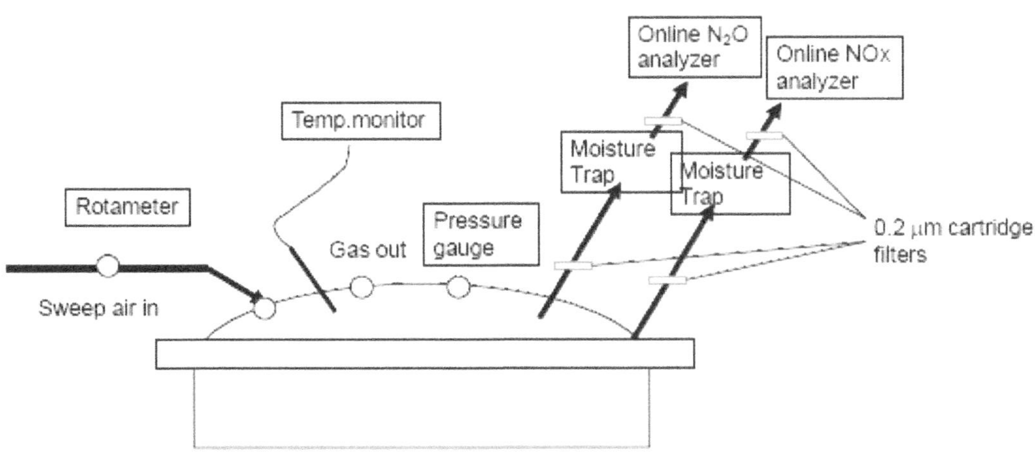

Figure 14.5 Schematic of Flux-chamber Employed for N_2O measurement.

The SEIFC consists of a floating enclosed space from which exhaust gas is collected in a real-time or discrete fashion. Since the surface area under the SEIFC can be measured, the specific flux of the gaseous compound of interest can be indirectly determined. The SEIFC 'floats' on the activated sludge tank surface and several replicate measurements can be taken at different locations in a single tank as well as from different tanks (nitrification, denitrification) along a treatment train.

The SEIFC is also equipped with mixing (physical mixer or via sweep gas circulation) to ensure adequate gas mixing and in some cases, an online temperature probe. The SEIFC is currently one of the few devices accepted by the USEPA for measuring gaseous fluxes (Tata et al., 2003) and as such will be employed for this study. Gas-phase analyses was conducted via infra-red (N_2O) and chemiluminescence (NO_x).

In general, sampling was conducted at multiple locations of the activated sludge train in each wastewater treatment facility. These locations included aerobic, anoxic and anaerobic zones, depending upon the configuration of the given facility. During the course of the gas phase sampling, liquid phase samples were collected adjacent to the hood location. The samples were filtered immediately upon collection in the field and analyzed by host plant personnel for ammonia, nitrite and nitrate concentration, utilizing readily available field methods (i.e. Hach kits) and standard laboratory analytical methods.

The specific locations selected were the geometric center of each demarcated anoxic or aerobic zone in the WWTP, or alternately locations where nitrification could be inferred based on initial screening of NH_4^+-N and DO concentrations (as in the plug-flow processes). For discrete measurement at each of these locations, 30 replicate measurements of gaseous N_2O and 1 measurement of aqueous N_2O were obtained over a period of 30 min. During continuous measurement at each of these specific locations over a 24 hour period, gaseous N_2O concentrations were still measured at 1/min, while aqueous N_2O concentrations were measured four-five times per day. Independent replication at each location (on different days) was not conducted owing to practical limitations associated with such an extensive campaign.

Sampling procedures-aqueous N_2O concentration

Aqueous phase N_2O concentrations were measured using a miniaturized Clark-type sensor with an internal reference and a guard cathode (Unisense, Aarhus, Denmark). The sensor is equipped with an oxygen front guard, which prevents oxygen from interfering with the N_2O measurements. The sensor is connected to a high-sensitivity picoammeter and the cathode is polarized against the internal reference. Driven by the external partial pressure, N_2O from the environment will penetrate through the sensor tip membranes and be reduced at the metal cathode surface. The picoammeter converts the resulting reduction current to a signal. The internal guard cathode is also polarized and scavenges oxygen in the electrolyte, thus minimizing zero-current and pre-polarization time.

Sampling procedures-measurement of advective gas flow rate from aerated zones

Advective flow of gas through the flux-chamber ($Q_{emission}$) in aerated zones was measured using a modification of ASTM method D1946. Briefly, a tracer gas consisting of 10,000 ppmv ($C_{helium-tracer}$) He was introduced into the flux-chamber at a known flow rate, Q_{tracer} (equation 1). He concentrations in the off-gas from the flux-chamber ($C_{helium-FC}$) were measured using a field gas-chromatograph equipped with a thermal conductivity detector (GC-TCD). $Q_{emission}$ was computed using equation 10.

$$Q_{tracer} * C_{helium-tracer} = (Q_{tracer} + Q_{emission}) * C_{helium-FC}$$
$$Q_{emission} = \frac{Q_{tracer} * (C_{helium-tracer} - C_{helium-FC})}{C_{helium-FC}} \quad (10)$$

Sampling procedures-measurement of advective gas flow rate from non aerated zones

The only modification to the protocol to measure the emission flow rate from *non-aerated zones* was the introduction of sweep gas (air) or carrier gas through the flux-chamber at a known flow rate (Q_{sweep}), in addition to the He tracer gas. The corresponding $Q_{emission}$ was computed using equation 11. Addition of sweep gas is needed to promote mixing of the SEIFC contents, owing to the low advective gas flow from the anoxic-zone headspace. Sweep-air N_2O concentrations were always measured and typically below the detection limits of the N_2O analyzer.

$$Q_{tracer} * C_{helium-tracer} = (Q_{tracer} + Q_{sweep} + Q_{emission}) * C_{helium-FC}$$
$$Q_{emission} = \frac{Q_{tracer} * (C_{helium-tracer} - C_{helium-FC})}{C_{helium-FC}} - Q_{sweep} \quad (11)$$

During continuous N_2O measurements, $Q_{emission}$ was determined several times a day.

Sampling Procedures-Wastewater and Process Characterization

Preliminary Data Gathering and Steady State Process Analysis.

The integral dependence of N_2O and NO emissions on the process operating conditions made the development of a preliminary reconnaissance analysis crucial. The following background information was collected from candidate evaluation sites:

- **Overall Plant Description.** First, general information related to treatment plant configuration, liquid and solids process flow diagrams, design criteria, major mechanical process equipment, etc from the plant's design reports and/or O&M manuals were obtained. In addition, the following secondary process operating data was gathered via meetings with plant operations teams and process engineers:
- **Secondary Process Configuration.** including zone configuration, zone volumes, operating set points, basins in service, aeration flow and distribution, recycle streams and flow rates (if applicable)
- **Plant Operating Data.** Summary of a minimum of three months plant data applicable to the treatment process to allow for characterization of the process influent, target and actual operating set points for key operational parameters (DO, SRT), effluent concentrations.

Intensive on-site sampling and analysis

Intensive diurnal sampling was conducted during each N_2O sampling campaign at the different plants to develop correlations between process state variables and gas-phase N_2O and NOx concentrations.

14.3.3 Calculation N_2O emission

Determination of fluxes

The net flux of gaseous N species (mg/min-m^2) was calculated based on the gas flow rate out of the flux chamber ($Q_{emission}$, L/min), gas concentration (parts per million) and the cross-sectional area of the SEIFC (m^2) (Equation 12).

$$\text{Flux} = \frac{Q_{\text{emission}} * C}{A} \tag{12}$$

Determination of lumped emission factors

The surface flux calculated from equation 12 was translated into the flux of a given zone by multiplying with the specific zone area. The N_2O emission fractions (mass/mass) for each WWTP at any given time point were computed by normalizing the measured flux from each zone in the facility to the daily influent total Kjeldahl nitrogen (TKN) loading according to equation 13. Emission fractions were averaged over the course of the diurnal sampling period and reported as the average (avg.) ± standard deviation (sd). for each individual process sampled.

During each campaign, wastewater nitrogen species concentrations including influent, bioreactor and effluent total Kjeldahl nitrogen (TKN), ammonium (NH_4^+-N), nitrite (NO_2^--N) and nitrate ($^-$-N) were measured simultaneously about six-times per day according to Standard Methods (Eaton et al., 2005) to supplement the gas-phase measurements. The discrete measurements were averaged to generate the emission fractions described in equation 13. Additionally, seven out of the twelve processes were sampled at minimum and maximum annual wastewater temperatures to examine seasonal temperature impacts on N_2O generation and emission.

$$Emission \quad fraction = \frac{\sum_{i=1}^{n} Flux_i * Area_i \quad (kg \quad N_2O - N)}{Daily \quad \inf luent \quad TKN \quad load \quad (kg - N)} \tag{13}$$

Where:

$Flux_i$ = N_2O emission flux calculated from the ith zone (kgN_2O-N·m^{-2}·d^{-1})

$Area_i$ = Surface area of the ith zone (m^2)

n = number of zones in a given facility from which N_2O fluxes are captured

Daily influent TKN load: Average influent load (influent flow rate * influent TKN concentrations) over the course of 24 hours.

On average, wastewater characterization was performed at each gas sampling location as well as in the tank influent and effluent about six times per day. At facilities where analysis was not as frequent, daily composite measurements were employed, for instance in the influent and effluent samples. Alternately, in some facilities, online measurement devices (for measuring pH, dissolved oxygen, redox potential and select N-species, including NH_4^+-N and $^-$-N) were employed at different locations of the activated sludge tank, which also facilitated the wastewater characterization efforts.

For aerobic zones, the helium-based advective gas-flow data were correlated to plant-recorded airflow rates for any given zone via linear regression and used to calculate diurnal N_2O emissions. For anoxic (non-aerated) zones lacking associated plant airflow data, the average of the experimentally obtained helium-based gas flow rates was used to calculate diurnal N_2O emissions.

Correlation between WWTP Operating Conditions and N_2O Emissions

Possible links between WWTP operating conditions and N_2O emission fluxes were examined via multivariate linear regression modeling of emissions and several wastewater state variables using SAS (Cary, NC). For aerobic zones, N_2O fluxes were correlated with liquid temperature and sampling location-specific pH, temperature, dissolved oxygen (DO), NH_4^+-N and NO_2^--N concentrations and multiplicative combinations thereof. For anoxic zones, N_2O fluxes were correlated with sampling location-specific soluble chemical oxygen demand (sCOD), pH, temperature, DO, NO_2^--N and $^-$-N concentrations and interactive combinations thereof. Assumptions of state variable normality and equal variance were evaluated using error residual and covariance plots (not shown). The variables not normally distributed were log-transformed, verified for normality and equal-variance and subsequently used for regression modeling. Time points where all state variables had not been measured simultaneously were not included in regression analysis.

14.3.4 Quality control

A detailed quality assurance project plan (QAPP) was prepared and submitted along with the protocol to the USEPA for review. The specifics of the QAPP and related procedures used were verified independently by a Project Quality Officer from Rutgers-The State University of New Jersey.

The validity of the measurements using the protocol developed for this study was determined via a parallel sampling effort among three teams on September 9 and 10, 2008 at one of the test wastewater treatment facilities. The plant staff used an USEPA flux chamber technology and sampled nitrogen oxide compounds using a field photo-acoustic analyzer. The Columbia University-WERF team used a flux chamber manufactured by St. Croix Sensory and measured N_2O off-gas concentrations via gas-filter correlation, described above. A third team led by Dr. Charles Schmidt used an USEPA flux chamber and sampled the off-gas into opaque Tedlar® bags for FTIR analysis (NIOSH 6660) by a commercial laboratory (Peak Analytical, Boulder, CO).

Based on these parallel measurements, the three methods resulted in similar results (Figure 14.6), with good correspondence in both the N_2O concentrations and off-gas flow-rate (only done by the Columbia-WERF team and Dr. Schmidt's team, Figure 14.7). Briefly, the following observations were made based on the results obtained and incorporated into subsequent full-scale measurement campaigns.

The use of an inert gas tracer was demonstrated to be an appropriate method to determine the advective off-gas flow rate. This was an improvement over the initial method developed by the Columbia-WERF team based on successive dilution of the N_2O concentrations. The initial method was discontinued following the validation study and replaced with He-tracer based method to determine advective flow rate.

Significant spatial and temporal variability in the measured concentrations of headspace N_2O was observed by the Columbia-WERF and Schmidt teams (Figure 14.8). Therefore, for subsequent full-scale measurements, discrete measurements (once a day) of N_2O at different locations in any given WWTP was discontinued. Rather, a substantially more involved sampling strategy that entailed 24 hour 'real-time online monitoring' of emissions at each location was initiated.

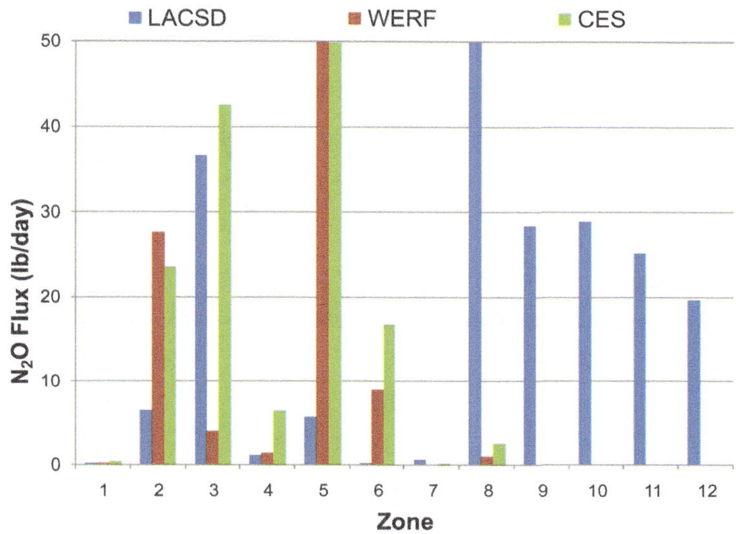

Figure 14.6 Comparison between N_2O fluxes obtained via three independent methods. Note, only the CES and WERF fluxes can be directly compared since they were conducted concurrently; (1 lb is equal to 0.45 kg)

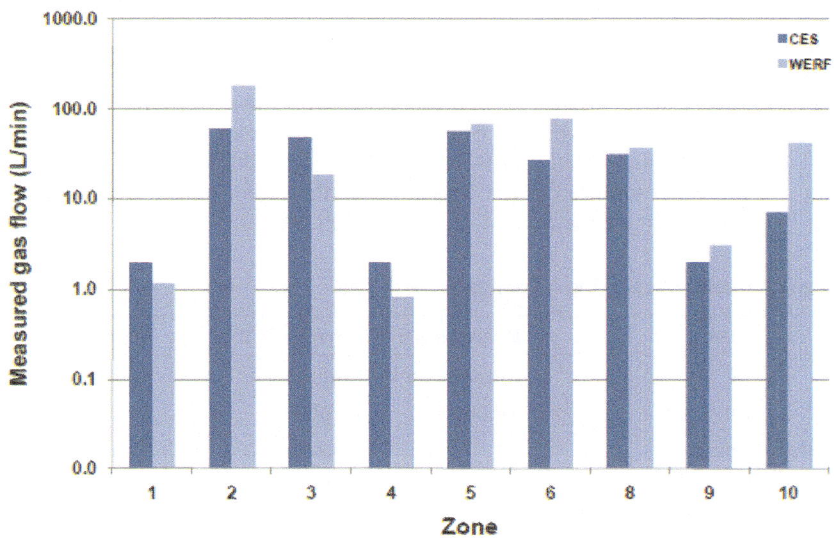

Figure 14.7 Comparison between gas flow rates obtained via the tracer gas (CES) and successive dilution (WERF-Columbia) methods

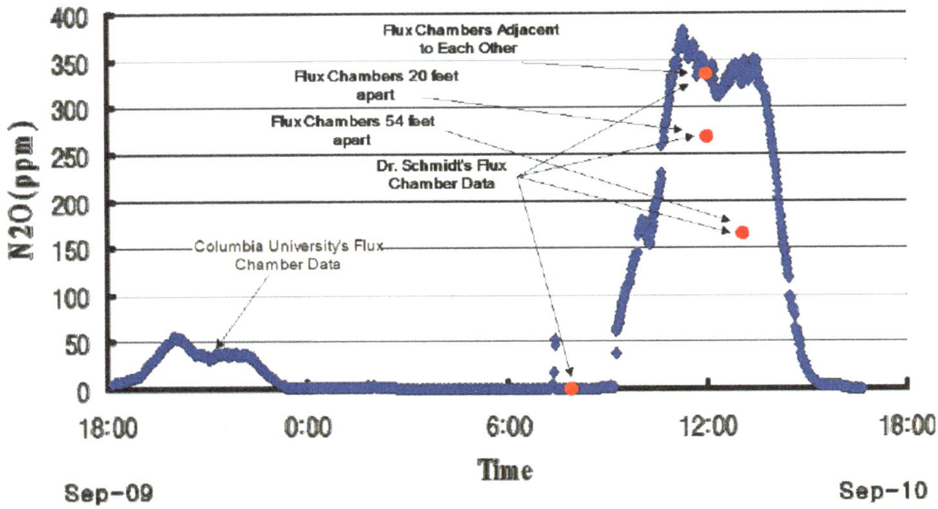

Figure 14.8 Illustration of spatial and temporal variability in N_2O concentrations in the headspace of an aerobic zone that necessitates real-time online monitoring. Columbia University's flux chamber data given by near continuous blue diamonds; Dr. Schmidt's flux chamber data given by 4 red dots.

14.4 NITROUS OXIDE MEASUREMENTS THE NETHERLANDS

14.4.1 Field sampling sites

Four measurement campaigns were performed in total. The WWTPs were selected to have covered activated sludge compartments. This allows for comparatively accurate knowledge on the gas flow. The measurement campaigns were done at two BNR carrousel activated sludge plants, Carrousel 1 (25–29th of September 2008), Carrousel 2 (29th May–4th of June 2009), and twice at a BNR (biological nitrogen removal) plant (14–21st of October 2008 and 9–16th of February 2009). The reason that at the BNR plant the emission of N_2O was measured twice was to study the effect of the temperature on the N_2O emission. In Table 14.4 the design characteristics of the WWTPs are shown.

Table 14.4 Design characteristics of the investigated WWTPs.

	Unit	Carrousel 1	Carrousel 2	BNR
Location	-	Papendrecht	Kortenoord	Kralingseveer
Hydraulic load	m^3/day	12,000	25,000	112,000
SRT (approximately)	days	23	unknown	20
Population equivalents	p.e. (54 gBZV)	40,000	100,000	360,000
Effluent requirements				
Nitrogen	mg/l	10	10	12

Phosphorus	mg/l	2	1	1.3
Organic load	kgCOD/day	3,540	13,880	25,000
Nitrogen load	kgN/day	410	1,200	3,700[1]
Organic sludge load	$kgCOD \cdot kg\ dw^{-1} \cdot d^{-1}$	0.031	0.060	0.048
Nitrogen sludge load	$kgN \cdot kg\ dw^{-1} \cdot d^{-1}$	0.009	0.012	0.024
Pre-sedimentation		No	No	Yes
P-removal		Biological	70% biological, 30% chemical	Biological
Type		carrousel	Carrousel	Pre-denitrification and –nitrification followed by two parallel carrousels
Sludge digestion		No	No	Yes

[1]Nitrogen load of the activated sludge units (after pre-sedimentation); the load includes nitrogen-rich internal flows from the sludge line

14.4.2 Samples collection and analysis

Nitrous oxide concentration

The emissions of N_2O, NO, O_2 and CO_2 were measured on-line using a custom-made analyzer (Rosemount, designed by the TU Delft). A switching-box equipped with four channels, controlled and measured four separate gas flows successively. The analyzer was set to take measurements every 60 seconds for a 180-second period, after which it would switch to the next channel, purge for 300 seconds and repeat. In each gas flow gaseous N_2O, NO, and CO_2 was measured by infra red, O_2 analysis was performed by para-magnetism (Emerson Process management Rosemount Analytical). All air flows through the analyzer were maintained at a flow of 1 L/min. Prior to passing through the analyzer all flows were led through condensers to prevent moisture from entering the analyzer.

At each WWTP the four channels were used to measure the N_2O concentration at different locations. An overview of the exact locations per WWTP is presented in Appendix 2.3. At all WWTPs one channel was used to measure the ambient air as a reference.

Additional grab-samples of off-gas from various (other) onsite processes were collected in gas sample bags. These were later analyzed for N_2O at the laboratory using gas chromatography (Varian CP-3800 Gas Chromatograph).

Gas flow

The flow of off gas through the pipes was measured using a pitot tube. The pitot tube was connected to a pressure probe (Testo 445, max. 10 hPa, accuracy +/- 0.03 hPa), which converted the pressure difference to a gas velocity. In this way the gas velocity was measured at least three times during the measuring campaign. During each measurement the gas velocity was measured at 20–50 individual points throughout the whole pipe. Based on the internal diameter of the pipe the gas flow was calculated. The pitot tube was used during the measurement campaigns in Papendrecht and Kralingseveer. During the measuring campaign at Kortenoord the gas velocity was measured with a hot wire anemometer (Testo 435–1, max. 20 m/s, accuracy +/- 0.03 m/s and +4% of measuring value). After the measurements the error of the measurement was determined. If this error was larger than the variation in the gasflow, the emission was calculated with an average gasflow, if the error was smaller than the emission was calculated with the daily gasflow.

Additional data

In addition to the gas measurements, concentrations of ammonium, nitrate, COD, BOD, Nkj and phosphate were obtained via daily analysis (24-hour flow-dependent sample collection) by the Waterboard. During the measurement campaign, field test kits were used as an indication of fluctuations of key parameters over the day (especially nitrite). Furthermore, on-line data of the available sensors at the WWTPs are logged as part of the general plant operation and were obtained from the Waterboards operating the plants. One parameter that was continuously logged was the influent flow. With this flow and the available analysis of the influent the nitrogen load could be calculated.

14.4.3 Calculation N_2O emission

The emission of N_2O is calculated based on the following measurements:

- N_2O concentration;
- gas flow;
- influent flow;
- NKj concentration (after primary settling including internal flow).

The total N_2O load was calculated, by averaging the N_2O concentration during the day. This average concentration was multiplied by an average gas flow. This could be done because it was found that the variation in the gas flow was smaller than the calculated error in the measurement.

The nitrogen load that entered the WWTP was calculated based on the measured NKj concentration and the average flow over the day.

14.4.4 Quality control

The N_2O analyser was two-point calibrated before every measuring campaign, using a standard gas for each individual gas (N_2O, NO, CO_2 and O_2). The accuracy of the analyser was 1%. The analyser was equipped with four measuring channels, one of them was always used to measure the ambient air. The concentrations in the ambient air were used to calculate the gas production (N_2O, NO, CO_2) or consumption (O_2). In this way it was also possible to neutralize a possible shift in the calibration line by per example a pressure difference.

For the quality control of the gas flow is referred to section 14.4.2.

14.5 METHANE MEASUREMENTS AUSTRALIA

14.5.1 Liquid phase measurement

While on-line sensors for measuring dissolved CH_4 are available, the measurement campaigns conducted in Australia so far measuring dissolved CH_4 in rising main sewers have solely relied on the use of manual sampling followed by laboratory analysis of the CH_4 content in wastewater samples using GC.

Field sampling sites

The design of the measurement campaign is important. As will be discussed later, the hydraulic retention time (HRT) of wastewater in a sewer line is a critical parameter influencing the production of CH_4. To obtain a representative picture of CH_4 formation in a given sewer line, it is important to design the sampling campaign so that a wide range of HRT is covered. This would involve taking wastewater samples at multiple locations along the sewer line at multiple times. Figure 14.9 shows the site map of the CO16 rising main sewer at the Gold Coast, Queensland, Australia, with the four sampling locations marked (Pumping Station, Sample Points 1, 2 and 3). In the sampling campaign reported in Foley et al. (2009), each of the sites was measured 3–4 times with an interval of 1 hour, which allowed to capture wastewater samples with HRT in the sewer line for 0–8.7 hours.

Figure 14.9 Aerial photograph of the CO16 rising main. Samples were collected from the CO16 Pump Station wet well (0 m), Sample point 1 (at 500 m), Sample point 2 (at 1100 m) and Sample point 3 (at 1900 m).

Sample collection and analysis

Wastewater sampling from **pressurized lines** was done through a special sampling arrangement (Figure 14.10). It consists of a 16 mm diameter pipe connecting a sampling tap at the ground level to the tapping arrangement attached to the underground pipe. Samples were collected using a hypodermic needle and 5 ml plastic syringe, attached directly to the pressurized rising main via a flexible hose. This procedure avoided any contact of the wastewater with atmosphere and possible oxygen interference.

Figure 14.10 Upper left and upper right: The sampling system. Lower: collection of wastewater samples containing dissolved CH_4.

Wastewater samples from **wet wells** were collected with a sampling device consisting of an open-head cylindrical container 40 cm in height and 1.5 l in volume. The sampler was submerged in the wet well for 5–10 seconds before its recovery. Samples were taken with 5 ml plastic syringe from a location 10 cm below the water surface in the sampler container to avoid contact with air.

The above method was also used for wastewater sampling from a **tapping point** (as an alternative method to what described in Figure 14.10). Prior to fill up the sampling container, sewage was thoroughly flushed for a period of one minute. Then, the hose-tube connected to the sampling tap was placed in the sampling container, which was filled from the bottom. Container was flushed for 30 seconds prior to a sample collection. Samples were taken 10 cm below the water surface.

A sample collected was subsequently injected into freshly vacuumed BD Vacutainer® tubes through a 0.22 μm pore diameter unit (Millipore, Millex GP). The Vacutainer tube was mixed overnight in a shaker to allow equilibration of gas and liquid phases. Most of the CH_4 (~97% at 25°C) would be transferred to the gas phase in this process (Alberto et al., 2000). The CH_4 concentration in the gas phase of the tube was measured using a Shimadzu GC-9A Gas Chromatograph equipped with a flame ionization detector (FID). The concentration of CH_4 in the initial liquid phase was then calculated using mass balance and

Henry's law (Guisasola et al., 2008). The Vacutainer tube was weighed before and after sampling to determine the sample volume collected. This volume, along with the known volume of the Vacutainer tube, enables to calculate dissolved CH_4 contained in the original wastewater sample.

14.5.2 Gas phase measurement

Due to safety considerations when operating in confined spaces, sewer workers in Australia commonly use portable gas detectors to detect toxic and flammable gasses. These sensors typically measure a number of gases including O_2, CO, H_2S, CO_2 and combustible gases such as CH_4. For CH_4 measurement, these sensors give direct readings of volumetric concentrations of CH_4 (%v/v) and/or percentages (%) of the lower explosive limit (LEL, which is approximately 5.1% v/v at 20°C). The detecting systems used include infrared sensors (IR), flam ionization detectors (FID) or catalytic bead sensors. These detectors are not specific for CH_4, and as such the measurement could be interfered by other hydrocarbons (e.g. ethane, propane, butane, etc). However, unless directly discharged, hydrocarbons other than CH_4 are not expected to be present in sewers in significant quantities. Therefore, the sensors should give correct readings of CH_4 concentrations in most cases despite of the possible cross interference. Automatic versions of the above gas detectors with data logging capabilities are also available for on-line CH_4 measurement. However, most of these sensors were developed for applications in other areas such as mining and landfills. Their application in the corrosive sewer environment is yet to be fully tested.

Methane concentration in sewer gas has also been measured through gas sampling (e.g. using Tedlar bags) followed by composition analysis using gas chromatography (GC). Compared to on-line sensors, this grab-sample based method could only give snapshots of CH_4 concentrations, which is a serious limitation. As will be discussed later, CH_4 concentration in both the liquid and gas phases is highly dynamic. It is difficult to capture the dynamics through offline sampling.

14.6 METHANE MEASUREMENTS FRANCE

The protocol for the CH_4 measurements in France is equal to the protocol used for the N_2O measurements as presented in section 14.2.

14.7 METHANE MEASUREMENTS USA

Work to date on the WERF (U2R08) project focuses on evolution of CH_4 from the collection system and from digester gas using equipment.

The research on the collection system has been performed in two phases, of which the first phase is completed and the second is underway. In the first phase the presence of CH_4 in the wet well was monitored instantaneously, in the second phase the presence of CH_4 was measured continuously (diurnal). In this report only the results of the first phase will be presented.

The work on the digester gas using equipment evaluated and estimated the fugitive CH_4 emissions from digester gas using equipment including Combined Heat and Power (CHP) technologies.

The work on the digester gas using equipment evaluated and estimated the fugitive CH_4 emissions from digester gas using equipment including Combined Heat and Power (CHP) technologies.

14.7.1 Collection system phase 1: Field sampling sites

The investigation on the collection system was performed in DeKalb County, Georgia. DeKalb County covers approximately 280 square miles (725 square km) and serves a population of over 700,000. In Appendix 2.8 a map of the County is provided and delineates the major drainage basins.

The map also identifies major wastewater collection system infrastructure. DeKalb's wastewater collection system is extensive, comprising of approximately 2,900 miles (1,931 km) of sanitary sewer, 80,000 manholes and 64 sanitary sewage lift stations[1]. Wastewater generated within the Snapfinger and Pole Bridge basins flows into one of DeKalb's two advance wastewater treatment facilities (AWTFs):

- Snapfinger Creek AWTF (36-mgd treatment capacity).
- Pole Bridge Creek AWTF (20-mgd treatment capacity)

A significant fraction of DeKalb's population lives within the Nancy Creek basin. Wastewater generated in the Nancy Creek basin is conveyed to City of Atlanta's R.M Clayton's WWTP via an 8 mile (12.87 km) long, 16 foot (4.88 m) diameter Nancy Creek Tunnel.

Of the County's 64 lift stations, 60 are small ranging in capacity from 80 to 700 gallons per minute (gpm) (0.3–2.6 m^3/minute) and 4 are large with firm capacities of over 2,000 gpm (~7.6 m^3/min) The small lift stations are not mechanically ventilated and have a gravity/passive ventilation.

The smaller lift stations comprise two general types of prefabricated units:

- Above-grade or Recessed. Includes a prefabricated concrete wetwell, four-level float controls, and two or four vacuum-primed pumps located directly above the wetwell (above grade) or in a recessed adjacent pit (recessed). Figure 14.11 shows a typical above-grade lift station.
- Below-grade. Has a separate prefabricated concrete or onsite constructed brick wetwell and prefabricated steel dry wells. Flooded-suction centrifugal pumps, motors, panel and switches are located below grade with ladder access.

A typical lift station in DeKalb County's selecting system is presented in Figure 14.11.

Figure 14.11 Typical lift station in DeKalb County's selecting system.

[1] The lift stations were predominantly fed by gravity sewers and in more limited cases by rising mains or a combination of the two.

Three of the large lift stations include Lower Crooked Creek Pumping Stations (LCC) 1, 2 and 3. These three lift stations operate in series serving to convey the bulk of the flow coming from Gwinnett County to the Pole Bridge Creek AWTF. They are very similar in design including their pumping capacities and wetwell dimensions. Each lift station has a wetwell and dry well with consant-speed pumps. The wetwells of LCC-1 and -2 are ventilated, whereas LCC-3 has an open, unventilated wetwell. Figure 14.12 shows a photo of LCC-2. Honey Creek Pumping Station is the fourth large pumping station and also has a dry well/wetwell configuration. Similar to LCC-3, the wetwell of this lift station is open to the atmosphere.

Figure 14.12 LCC-PS-2.

14.7.2 Collection system phase 1: Time periods

In order to study the effects of seasonal variations and temperature changes on the CH_4 emissions CH_4 monitoring was conducted during two discrete events: once during cold weather (winter monitoring) and once during warmer weather (summer monitoring).

The year was divided into 'cold' and 'warm' weather months. Cold weather months were selected as those with monthly average temperatures lower than the annual average temperature. Similarly, warm weather months were selected as those with monthly average temperatures higher than the annual average temperature. The selected cold and warm weather periods are depicted in Figure 14.12 together with the temperature profile of DeKalb County for the year 2009.

14.7.3 Collection system phase 1: Sample collection and analysis

Instantaneous CH_4 readings from each of the 64 lift stations were conducted as part of this phase. In order to better understand the chemistry and biological pathways, other parameters such as temperature, H_2S, CO_2 and CO content were also collected.

Sample collection in unventilated lift stations

Figure 14.13 shows the inside of a typical unventilated lift station in the DeKalb collection system. Methane concentrations were measured at three locations in each wetwell during each sampling event: immediately above the liquid surface, mid-way up the wetwell headspace; and immediately below the access hatch. While all three concentrations were collected, the highest of the three was used in order to somewhat mitigate the lack of accounting for dispersion of lighter-than-air CH_4 in between pumping cycles or the air dilution that results when air is pulled into the well during a pumping cycle. A similar procedure for data collection was used at large unventilated lift stations with open forebays including LCC PS-3 and Honey Creek.

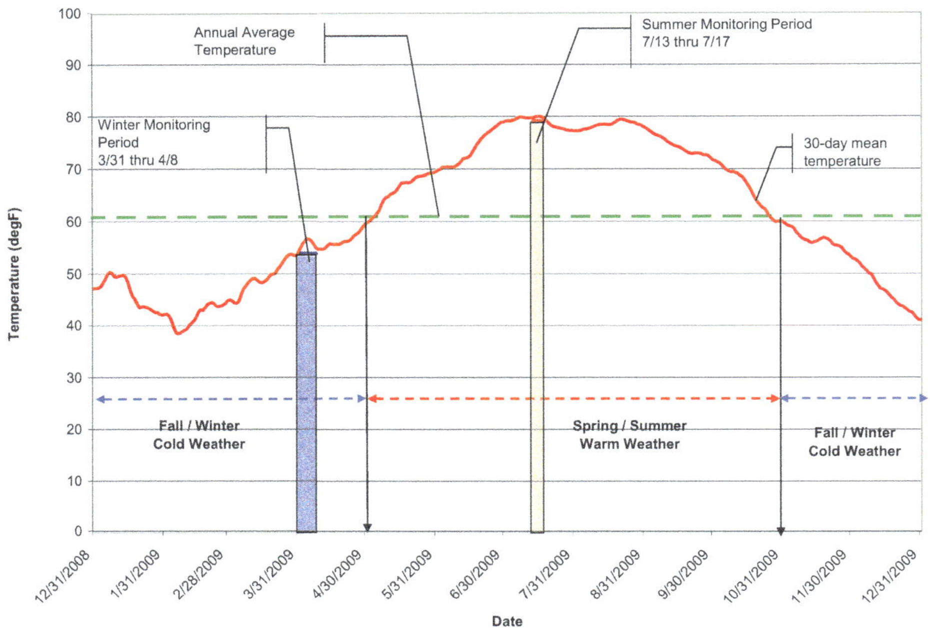

Figure 14.13 Temperature profile of DeKalb County for the year 2009 and the sampling periods.

Figure 14.14 Unventilated wetwell.

Sample collection in ventilated lift stations

There is only one ventilated lift station within DeKalb's collection system. At both LCC-1 the wetwell is covered and foul air exhaust fans continuously withdraw air from and send it to an odor control systems. Openings for the influent screens and at the wetwell covers allow air to enter the wetwell as foul air is exhausted. Instantaneous measurements for CH_4 and other parameters were sampled directly from the discharge header of the exhaust fan upstream of the biofilter.

Analysis

Portable instruments were used for taking instantaneous readings in the field. A hand-held flame ionization detector (FID, MicroFID manufactured by PhotoVac) with hydrogen as a fuel source was used to measure CH_4. A portable four-gas analyzer with photoionization detector (PID, RKI Eagle Gas Portable Monitor) was used to measure H_2S, CO_2, %lower explosive limit (LEL) and CO. Ambient air and raw sewage temperatures were also monitored to check for a correlation with the measured CH_4 concentration. A dissolved oxygen (DO) probe (YSI 550A DO) was used to measure ambient air temperature, raw sewage temperature, and dissolved oxygen concentration in the raw sewage.

Gas flow in unventilated lift stations

The operation of constant-speed unventilated wetwells creates synchronized 'breathing cycles' in the wetwell headspace. As the water level in a wetwell rises, the air inside the wetwell volume is expelled into the atmosphere through the vent or the wetwell cover. Conversely, when the wetwell water level is drawn down by pumping, outside air is drawn into the wetwell headspace.

Since all of the wetwells are at or near the atmospheric pressure, the amount of air breathed in and exhausted out during each pumping cycle is approximately equal to the active volume inside the wetwell or the volume of raw sewage pumped during each cycle. Thus, the total volume of air exhausted out of a given lift station is assumed equal to the gallons of raw sewage pumped over the same period.

During the separate DeKalb collection systems surveying effort, the active volume of each lift station was determined along with the pumping rate for each of the pumps in all 61 lift stations. In order to keep track of the monthly flow rate of the lift stations, DeKalb has installed pump-hour meters at each lift station and maintains monthly logs of hour-meter readings. These hour-meter logs along with the pumping capacities established during the collection systems study were used to determine the monthly flow rates for each of the lift stations. Using this method, monthly flows for the one-year period from September 2008 to August 2009 (the 'model period') were established for each of the smaller constant-speed lift stations.

For each of the unventilated lift stations, CH_4 readings from cold weather monitoring (March/April) were applied to monthly flows from November through April and CH_4 readings from warm weather monitoring were applied to monthly flows from May through October to estimate the annual CH_4 emissions.

While this approach allows an annual mass emission to be calculated for the DeKalb pumping stations, it is likely subject to be a number of limitations, namely:

- Dispersion of CH_4, which is lighter than air, is not accounted for. Many stations cycle only a few times per day and those stations likely lose a considerable portion of the evolved CH_4 to the atmosphere in between cycles. This phenomenon would result in under-reporting of actual emissions.
- Dispersion effects can be very significant at some of the larger pumping stations, like LCC-3. LCC-3 has a large, uncovered forebay with significant interchanges of outside air. On calm days the forbay air changes could be 1 to 5 times per hour and could increase to as much as 10 or 30 times per hour on windy days. This limitation, combined with the fact that these stations have

the highest volume and mass throughput which should produce more CH_4, could represent a very significant source for under-reporting.

- During a pumping cycle, a significant volume of outside air is drawn into the wet well. If a sampling event occurred soon after such a dilution, the applied concentrations could under-report the actual CH_4 emissions.
- All of the sampling occurred during daylight hours and as such, correlated with higher load periods during the day that would be thought to coincide with periods of higher CH_4 production. This limitation could result in over-reporting of actual emissions.

Gas flow in ventilated lift stations

In order to compute the mass of CH_4 emissions, exhaust fan air flow rate and CH_4 concentration measured in the discharge header were used. In the absence of measured air flow data, the foul air fan flow rate was determined to be 1,950 cfm (cubic foot per minute) using the manufacturer's fan curve and field-observed pressure conditions. The instantaneously-measured winter and summer CH_4 concentrations were applied to the 1,950 cfm in order to calculate the mass emissions from this station.

It is possible that this approach over reports the CH_4 emissions from LCC-1 in that the sampling occurred during higher flow periods when more flow and mass would be thought to produce higher-than-average CH_4 emissions.

14.8 METHANE MEASUREMENTS THE NETHERLANDS

14.8.1 Field sampling sites

The emission of CH_4 was determined at the same WWTPs as for the N_2O emission. The WWTPs at Papendrecht and Kortenoord do not have a sludge digester; one is present at WWTP Kralingseveer. Compared to the N_2O emission the emission of CH_4 was not measured continuously, but was measured via grab samples.

14.8.2 Samples collection and analysis

Sample collection

Grab samples for the analysis of the CH_4 concentration were taken at the locations were emission of CH_4 could be expected. The amount of grab samples taken at each WWTP is summarized in Table 14.5.

Table 14.5 Summary locations CH_4 measurements and number of samples taken.

WWTP	Sample location	Number of samples
Papendrecht	• After inlet work and coarse screen	3
	• Selector	2
	• Anaerobic tank	2
	• Sludge loading	1
	• Carrousels	1
Kortenoord	• After inlet work and coarse screen and grid removal	5
	• Selector	5

	• Aeration tank (3 channels N_2O measurements)	3	
	• Sludge thickeners	3	
	• Sludge storage	3	
	• Sludge dewatering site	4	
Kralingseveer		October	February
	• After inlet work and coarse screen	2	3
	• Primary settling	2	3
	• Selector	2	3
	• Sludge thickeners	2	3
	• Sludge storage	2	3
	• Aeration tank	2	2
	• Carrousel 1 a.nd 2	2	3

Analysis

The grab samples were analyzed for CH_4 at the laboratory using gas chromatography (Varian CP-3800 Gas Chromatograph).

Gas flow

The gas flow was determined in the same way as for the emission of N_2O (see section 14.4.2).

Calculations

The emission of CH_4 is based on the following measurements:

- CH_4 concentration in grab samples;
- Gas flow;
- Influent flow;
- COD concentration influent (after primary settling, including internal flow).

The CH_4 emission was calculated based on the average concentration in the grab samples. In almost all cases[2] the average concentration was multiplied by the average gas flow as the error in the measurements was larger than the variations in the gas flow.

14.9 TOTAL CARBON FOOTPRINT WWTP

In the research performed in the Netherlands the emission of N_2O and CH_4 was measured at the same three WWTPs. Based on these measurements it was possible to determine the contribution of N_2O and CH_4 to the total carbon footprint of a WWTP including the use of electricity and natural gas. To determine the total carbon footprint of a WWTP all sources were converted to CO_2 equivalents. The conversions factors that have been used in this case are summarised in Table 14.6.

[2] Exceptions were selector Papendrecht, sludge storage in Kralingseveer (February), here the daily gas flow was used.

Table 14.6 Conversion factors for climate footprint calculations used in the Netherlands.

	Conversion factor	**Unit**
Electricity	0.67	kg CO_2/kWh
Natural gas	1.8	kg CO_2/Nm^3
N_2O	298[1]	kg CO_2/kg N_2O
CH_4	25[1]	kg CO_2/ kg CH_4

[1]GWP AR4: Adjusted GWP in IPCC Fourth assessment report, 2007 (IPCC, 2007).

The conversion factors mentioned for electricity and natural gas in Table 14.6 are specific for the situation in the Netherlands, for any other country other factors do apply.

Chapter 15
Results

15.1 NITROUS OXIDE EMISSION AUSTRALIA

15.1.1 Emission and origin

Figure 15.1 shows the dissolved N_2O concentrations (and associated 95% confidence intervals) measured by the microsensor in each reactor zone, at each WWTP. A wide range of values is apparent across the reactor zones, sampling rounds and WWTPs, from near zero to greater than 1 mg/l.

Figure 15.2 shows the normalised net N_2O-N generation (GF_R) and mass transfer emission profiles (EF_R), per reactor zone and sampling round, at each of the continuous flow WWTPs.

The calculated total WWTP generation factors (GF_{WWTP}) for the 20 sampling rounds are reported in Table 15.1 and show a wide range of results across two orders of magnitude. The minimum generation factor was 0.002, and the maximum 0.112 $kgN_2O\text{-}N/kgTKN_{influent}$. The average of the 20 samples was 0.016 ± 0.012 (t-dist, $\alpha = 0.05$). However, close inspection of these results shows the average to be skewed upwards by four results greater than 0.02 kgN_2O-N per $kgTKN_{influent}$. Excluding these results from the statistical sample, the average of the remaining 16 results is 0.006 ± 0.002 $kgN_2O\text{-}N/kgTKN_{influent}$.

Table 15.1 also reports the percentage of generated N_2O-N that is lost via mass transfer to the atmosphere. Generated N_2O-N can also be lost as dissolved N_2O-N in the WWTP effluent and waste solids. However, it is clear from the results in Table 15.1, that these losses are generally minor (i.e. <5%). Whilst N_2O is a highly soluble gas in water, it also has a high mass transfer coefficient (similar to oxygen), so it is not unexpected that the majority of N_2O generated in the bioreactors is quickly stripped to the atmosphere.

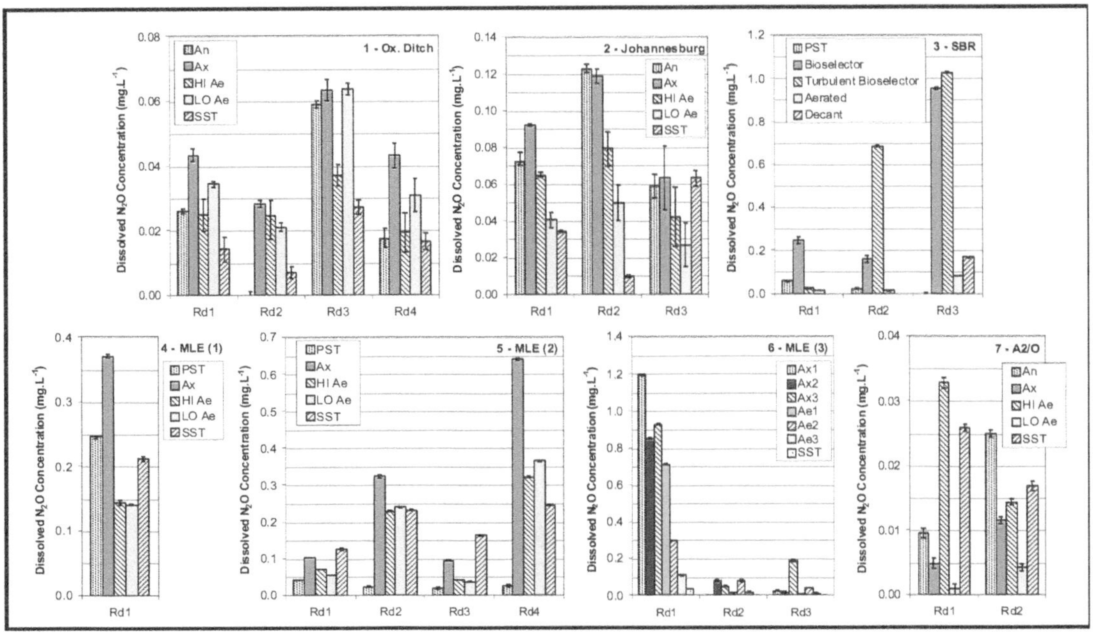

Figure 15.1 Dissolved N_2O concentrations measured at each WWTP. Error bars indicate the 95% confidence interval, based on the microsensor sampling data.

Shown in Figure 15.2 are the N_2O-N generation factor results plotted against bulk bioreactor NO_2^--N concentration, and two process design parameters, namely effluent total nitrogen and a-recycle rate (i.e. recycle rate from aerobic zone to anoxic zone). Figure 15.2A suggests that the high N_2O-N generation factors generally correspond with high bulk NO_2^--N concentrations in the bioreactor. Figure 15.2B and Figure 15.2C together show particular design parameters that potentially influence the stability of WWTPs in relation to N_2O generation.

The predominance of mass transfer emissions from the aerated zones is in agreement with other full-scale studies of aerated and quiescent bioreactors (Czepiel et al., 1995).

The profiles in Figure 15.1 show that whilst little atmospheric emission occurs from the quiescent anoxic zones, they actually are responsible for a substantial portion of the N_2O generated in the bioreactor. Examination of the net generation profile, rather than the emission profile, is therefore more useful in the context of understanding (and possibly controlling) the underlying mechanisms of N_2O formation.

The measurement of positive dissolved N_2O concentrations in all reactor zones across all seven WWTPs (refer to Figure 15.1) suggests that N_2O is pervasive in most BNR plants. Furthermore, the calculation of positive N_2O-N generation factors for all WWTPs in Table 15.1 confirms that these facilities are sources of anthropogenic greenhouse gas emissions for UNFCCC accounting purposes. It should also be noted that the generation factors for all sample rounds were positive, even when accounting for the uncertainty of a best-case combination of a low mass transfer coefficient and low dissolved N_2O concentration.

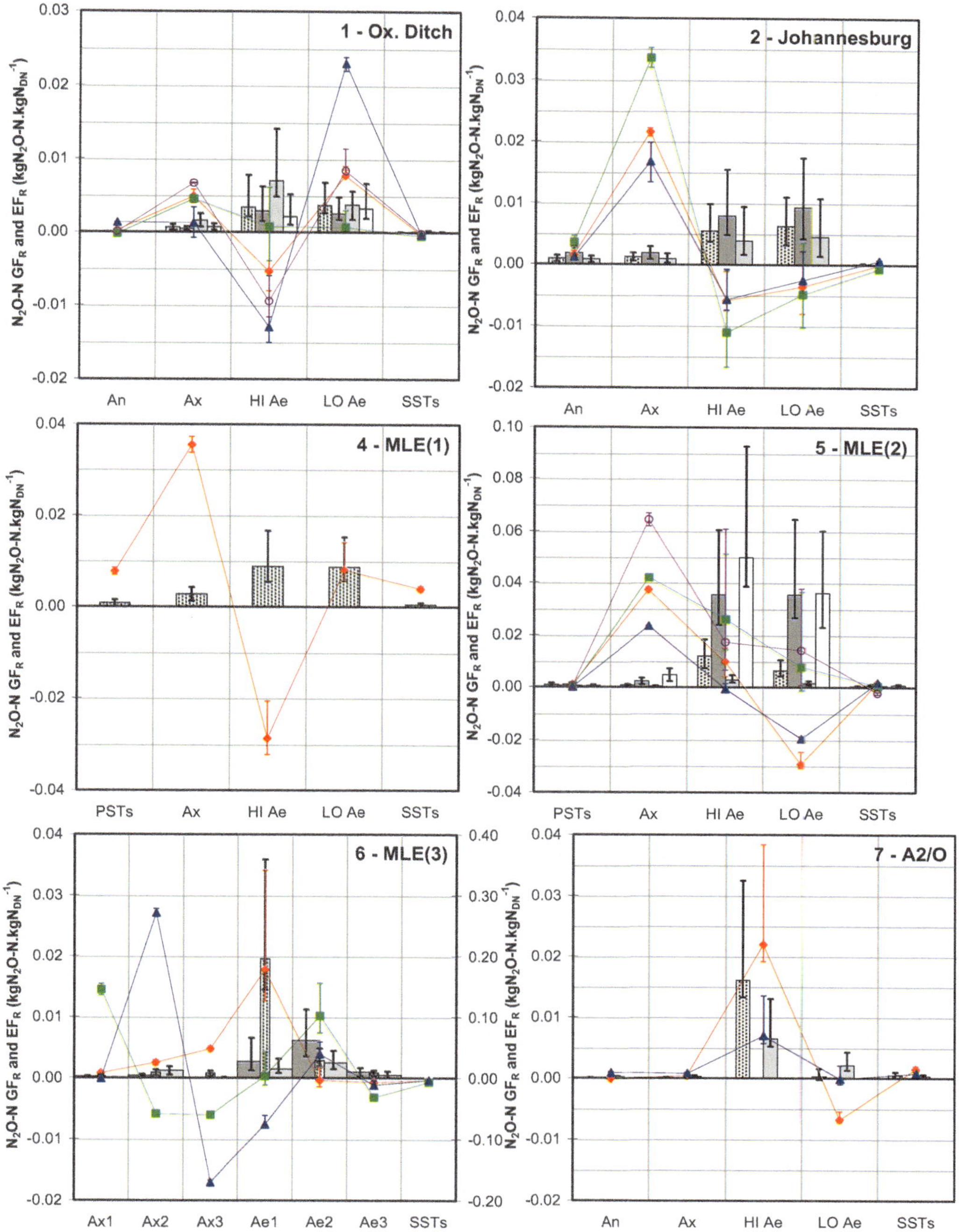
1 - Ox. Ditch
2 - Johannesburg
4 - MLE(1)
5 - MLE(2)
6 - MLE(3)
7 - A2/O
N_2O-N GF_R and EF_R (kgN_2O-$N.kgN_{DN}^{-1}$)
An Ax HI Ae LO Ae SSTs
PSTs Ax HI Ae LO Ae SSTs
Ax1 Ax2 Ax3 Ae1 Ae2 Ae3 SSTs
0.03 0.02 0.01 0.00 -0.01 -0.02
0.04 0.03 0.02 0.01 0.00 -0.01 -0.02
0.04 0.02 0.00 -0.02 -0.04
0.10 0.08 0.06 0.04 0.02 0.00 -0.02 -0.04
0.40 0.30 0.20 0.10 0.00 -0.10 -0.20

Figure 15.2 Net N_2O-N generation (GF_R) and mass transfer emissions (EF_R) profiles, per reactor zone, in each sampling round at the six continuous flow WWTPs. "An" – anaerobic zone, "Ax" – anoxic zone, "HI Ae" – highly aerated aerobic zone, "LO Ae" – less aerated aerobic zone, "PSTs" – primary sedimentation tanks, "SSTs" – secondary sedimentation tanks. Results from Round 1 at WWTP No. 6 are plotted on the secondary axis – note the change in scale. Error bars indicate the combined uncertainty from the 95% confidence intervals of both kLa and measured dissolved N_2O concentrations.

The variation in generation factor results in Table 15.1 highlights two key points. Firstly, WWTP Nos. 1 and 2 had low generation factors, with low variability. This was true also for the limited sampling rounds at WWTP No.7. Secondly, WWTP Nos. 3, 5 and 6 generally had higher generation factors, but also exhibited significantly higher variability over a short time frame. For example at WWTP No.5, the generation factor varied from 0.009 $kgN_2O\text{-}N/kgTKN_{influent}$ in the morning sampling round, to 0.030 in the afternoon sampling, decreasing to 0.004 the following morning, and then increased again to 0.035 that afternoon. The results also highlight that some WWTPs might operate steadily with relatively low N_2O generation, and then suffer some process perturbation that leads to a temporary spike in N_2O formation[1]. This was witnessed most noticeably at WWTP No.6, which had a peak N_2O GF of 0.112 kgN_2O-N per $kgTKN_{influent}$ in the first sampling round, but relatively low GFs in the remaining sampling rounds. Together, these two points indicate that both process design and variability in operating process conditions are likely to influence the magnitude and variability of N_2O-N generation factors.

15.1.2 Process parameters of influence

Figure 15.3A suggests that high N_2O-N generation factors generally correspond with higher bulk **NO_2^--N** concentrations in the bioreactor. Although the number of data points is limited, there appears to be a threshold value at approximately 0.3 - 0.5 $mg.l^{-1}$ NO_2^--N at which the generation factor jumps to be >> 0.03 $kgN_2O\text{-}N/kgN_{denitrified}$. All four results > 0.03 $kgN_2O\text{-}N/kgN_{denitrified}$ occurred under conditions where bulk NO_2^--N concentrations exceeded 0.4 mg/l.

Figure 15.3B and Figure 15.3C together show particular design parameters that potentially influence the stability of WWTPs in relation to N_2O generation. Both WWTP Nos. 1 and 2 have very high a-recycle rates and correspondingly low effluent TN concentrations. The WWTPs that had more highly variable N_2O-N generation factors tended to have lower a–recycle rates and correspondingly higher effluent TN concentrations. From Figure 15.3B, there appears to be a threshold at approximately TN 10 mg/l, below which the N_2O-N generation factor is relatively low and more stable.

[1] That a temporary peak in N_2O emission is possible as a result of a process perturbation was also found by Burgess (2002). In this case a peak of N_2O was found after a peak in NH_4 or a disturbance in the aeration.

Table 15.1 Nitrous oxide generation factor (GF_{WWTP}) and % mass transfer emissions ($\%Tr_{WWTP}$) for each WWTP and sample round.

WWTP No.	N_2O-N GF_{WWTP} (kg N_2O-N/kgTKN$_{influent}$) Note 2,3				N_2O-N Mass Transfer Emissions ($\%Tr_{WWTP}$)			
	Round 1	Round 2	Round 3	Round 4	Round 1	Round 2	Round 3	Round 4
1– Ox. Ditch	0.003 (0.002–0.006)	0.002 (0.001–0.004)	0.004 (0.002–0.007)	0.002 (0.001–0.005)	98.2–99.0%	98.5–99.3%	95.3–98.2%	96.3–98.7%
2– Johannesburg	0.010 (0.006–0.017)	0.017 (0.009–0.031)	0.007 (0.003–0.017)		96.8–98.8%	99.0–99.6%	88.4–97.7%	
3– SBR	0.006 (0.005–0.011)	0.007 (0.006–0.014)	**0.032** [1] (0.029–0.065)		94.6–97.4%	94.6–97.5%	93.7–97.2%	
4– MLE (1)	0.012 (0.008–0.020)				67.7–86.3%			
5– MLE (2)	0.009 (0.006–0.015)	**0.030** [1] (0.021–0.051)	0.004 (0.002–0.005)	**0.035** [1] (0.025–0.061)	88.3–95.3%	94.1–97.6%	71.4–87.4%	94.5–97.8%
6– MLE (3)	**0.112** [1] (0.078–0.200)	0.005 (0.003–0.010)	0.003 (0.002–0.006)		99.6–99.8%	99.7–99.9%	99.9%	
7– A^2O	0.007 (0.006–0.015)		0.004 (0.003–0.008)		95.1–98.0%		94.0–97.6%	

Notes:
[1] **Bold** results are greater than 0.02 kgN_2O-N.kgTKN$^{-1}_{influent}$.
[2] For the purposes of this report, the units of GF_{WWTP} have been converted from kg N_2O-N.kgN$^{-1}_{denitrified}$ to kg N_2O-N.kgTKN$^{-1}_{influent}$.
[3] Figures in brackets represent the uncertainty range for GF_{WWTP} results, based on the combination of 95% confidence intervals for both k_La values and measured dissolved N_2O concentrations.

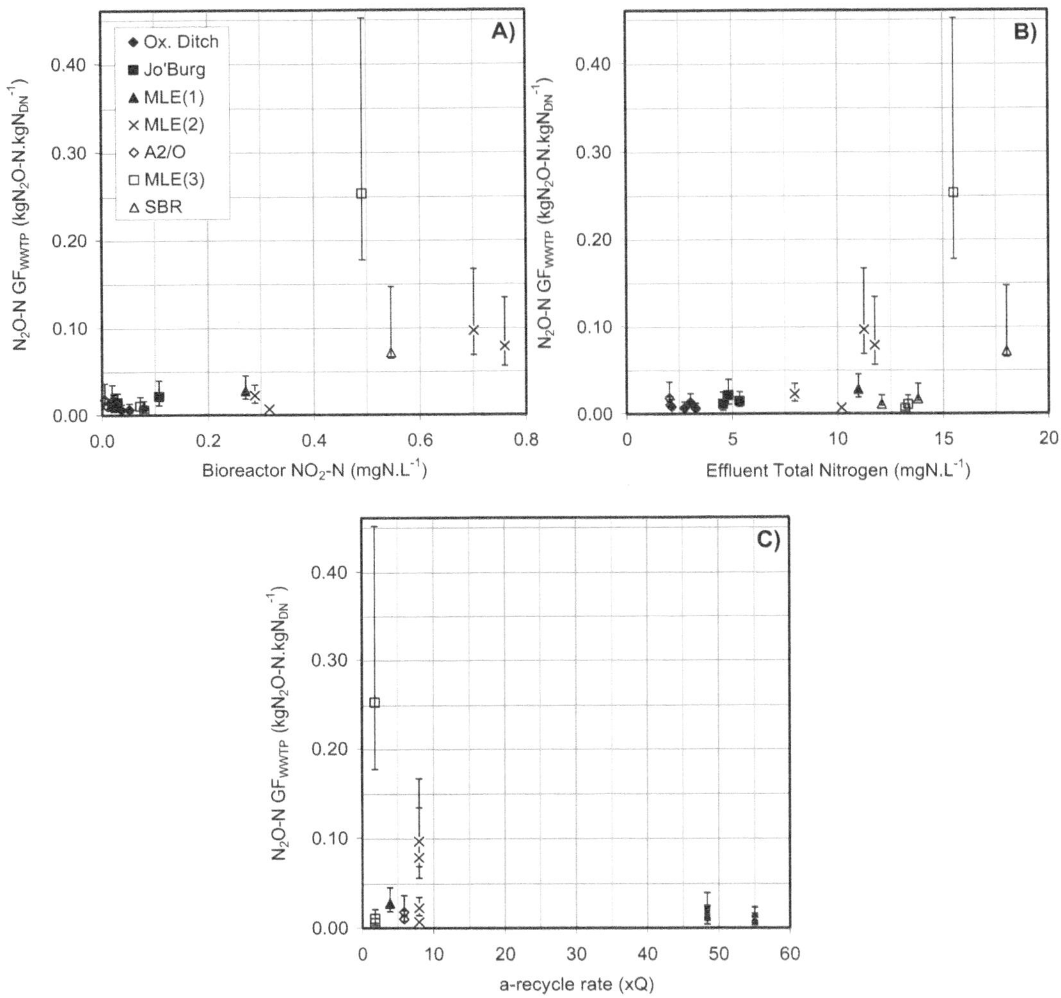

Figure 15.3 Net N_2O-N generation factor, GF_{WWTP}, in each sampling round, plotted against A) bulk bioreactor nitrite-N concentration, B) effluent total nitrogen, and C) a-recycle rate (aerobic → anoxic) as a multiple of the average influent flowrate, Q.

15.2 NITROUS OXIDE EMISSION FRANCE

15.2.1 Emission of N_2O

The emission of N_2O that was measured at the four WWTPs is presented in Table 15.2 .

Table 15.2 Emission of N_2O at four WWTPs in France.

WWTP	Emission (kg N_2O/kg $TKN_{influent}$)	Emission (kg N_2O-N/kg $TKN_{influent}$)
WWTP 1	<0.0002	<0.0001
WWTP 2	<0.0002	<0.0001
WWTP 3	0.003	0.0017
WWTP 4	0.002	0.0011

15.2.2 Origin N_2O emission

In all cases N_2O emissions were emitted in aerated zones. Emissions profiles show that some N_2O can be produced during non aerated (anoxic) steps but it is always emitted during aerated steps (stripping). This is illustrated with the results from WWTP 1, 3 and 4.

WWTP 1, CSTR

Emissions of N_2O during nitrogen removal were investigated in a completely stirred tank reactor (activated sludge) at full scale WWTP (WWTP1). The emission of N_2O around the start of the aeration is presented in Figure 15.4.

From Figure 15.4 the following can be observed:

- A first peak of N_2O at the start of the aeration. This is probably due to stripping of N_2O that was previously generated.
- A second peak of emission is observed during the nitrification phase. As can be observed N_2O emissions are linked to nitrification. Dissolved O_2 was around 2 mg/l during the nitrification and increased to 6 mg/l when all NH_4^+-N has been oxidized (no data on the NH_4 concentration in the basin).

WWTP 3, plug flow reactor

Emissions of N_2O during nitrogen removal were investigated in a plug flow activated sludge tank at a full scale WWTP (WWTP 3). Nitrous oxide was sampled at nine points (see Figure 15.5) with the square wooden sampling box as described in 5.2.2, following a 'kinetic' approach. All N forms were simultaneously monitored in water (NH_4^+-N, NO_2^--N, $^-$-N, and dissolved N_2O) and in air (N_2O).

The N_2O generation and emission during the aerated step in a plug flow system is presented in Figure 15.6. The first 80 metres of the basin is an anoxic zone where denitrification occurs. No N_2O emissions have been observed during denitrification. In the aerated length (80-290m) NH_4^+-N and $^-$-N concentrations account for nitrification (NH_4^+-N oxidation). The influent concentration was 17 mg NH_4^+-N/l without nitrate neither dissolved N_2O. During nitrification NH_4^+-N has been removed while the nitrate concentration has increased to 15 mg $^-$-N/l. Dissolved N_2O was measured and the curve indicates that N_2O is generated during the nitrification step. As shown on the upper part of the graph N_2O emissions were observed and have been quantified. Dissolved oxygen was 0.5-1mg/l and increased to 5 mg/l when nitrification has been completed –i.e. when overall ammonium has been oxidized.

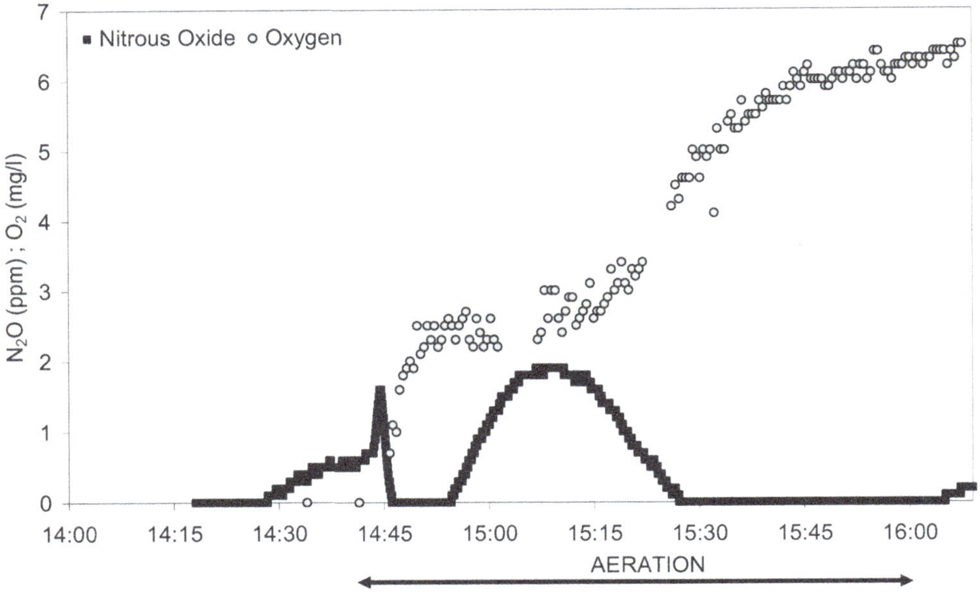

Figure 15.4 N_2O emission during nitrification – 06/2007.

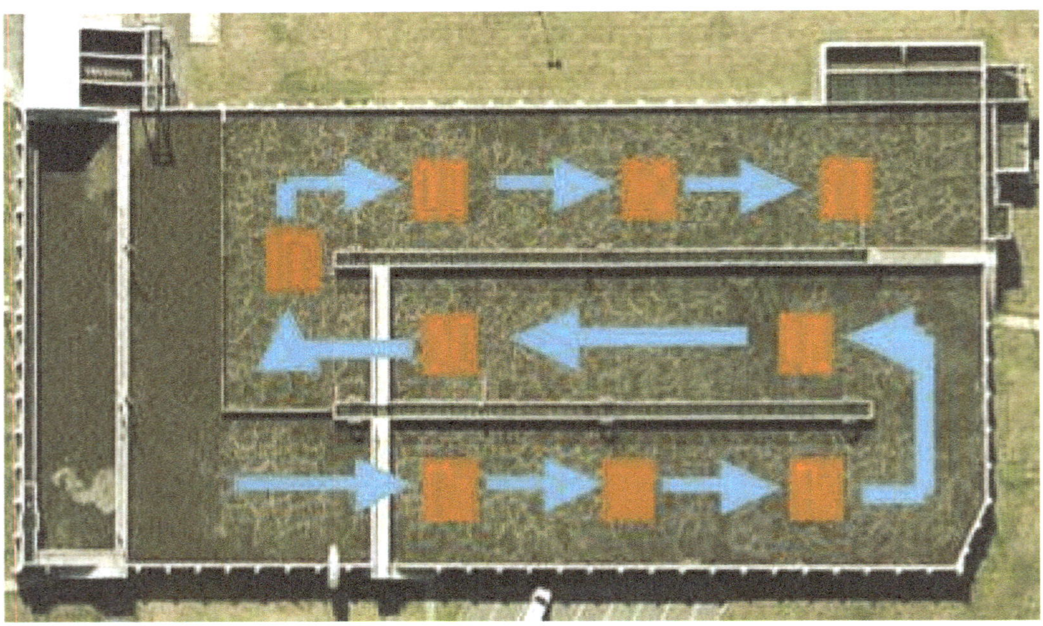

Figure 15.5 Sampling points for N_2O measurements at WWTP 3 in France.

WWTP 4, activated sludge with sequenced aeration followed by a MBR,

In this WWTP it was observed that:

- At the start of the aeration a first peak of N_2O appeared.
- A second peak of N_2O appeared during the course of the aeration period.

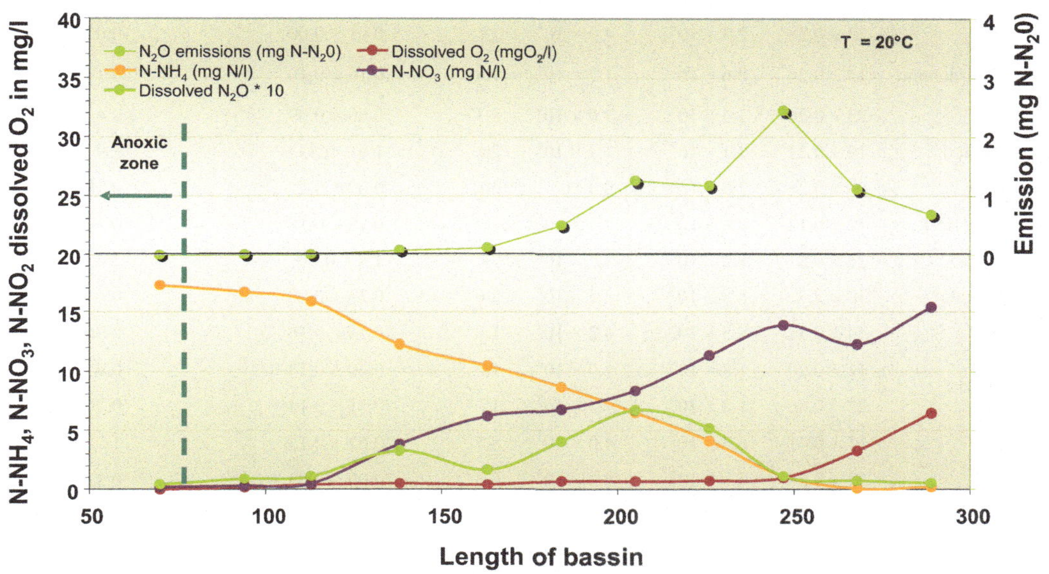

Figure 15.6 Emission of N_2O during nitrification in a full scale plug flow activated sludge tank at WWTP 3.

15.2.3 Process parameters of influence

Based on the work in France the following process parameters of influence have been identified: NH_4^+-N and DO. Furthermore air-flow must also be considered regarding stripping of N_2O.

15.3 NITROUS OXIDE EMISSION UNITED STATES OF AMERICA

15.3.1 Emission of N_2O

A wide range of N_2O emissions was measured across the twelve WWTPs operated at different temperatures, configurations and influent characteristics (Table 15.3).

Table 15.3 Summary of N_2O fluxes and emission factors measured at full-scale WWTPs.[2]

Plant Configuration	Temp(°C)	Reactor influent TKN load (g-N/day)	Reactor effluent TN load (g-N/day)	Q (MGD)	kg N_2O-N / kg $N_{influent}$ (%)	kg N_2O-N / kg $N_{denitrified}$ (%)
Separate-stage BNR	15 ± 0.48	1.8×10^6	3.6×10^5	23	0.03 ± 0.00	0.03 ± 0.01
	23 ± 0.28	2.3×10^6	4.3×10^5	27	0.01 ± 0.00	0.01 ± 0.00
Four-stage Bardenpho	14 ± 0.26	8.6×10^5	1.7×10^5	7.8	0.16 ± 0.10	0.19 ± 0.12
	23 ± 0.20	7.4×10^5	7.6×10^4	8.1	0.60 ± 0.29	0.66 ± 0.32
Step-feed BNR 1	19 ± 0.22	3.1×10^6	1.4×10^6	29	1.6 ± 0.83	2.9 ± 1.5
	25 ± 0.28	2.9×10^6	9.4 10^5	30	0.62 ± 0.27	0.90 ± 0.39
Step-feed non-BNR	17 ± 0.12	8.6×10^6	4.4×10^6	71	0.18 ± 0.18	0.37 ± 0.36
	26 ± 0.81	8.9×10^6	4.2×10^6	93	1.8 ± 0.79	3.3 ± 1.5
Separate centrate*	30 ± 2.3	8.8×10^6	5.5×10^6	2.0	0.24 ± 0.02	0.63 ± 0.06
	34 ± 0.32	8.5×10^6	4.2×10^6	1.6	0.54 ± 0.16	0.96 ± 0.32
Plug-flow 1	11 ± 0.20	1.8×10^6	1.0×10^6	18	0.40 ± 0.14	0.92 ± 0.32
	23 ± 0.46	1.8×10^6	7.3×10^5	15	0.41 ± 0.14	0.70 ± 0.24
Plug-flow 2	11 ± 0.41	6.3×10^5	4.0×10^5	8.7	0.62 ± 0.15	1.7 ± 0.41
	22 ± 0.58	6.6×10^5	4.0×10^5	6.6	0.09 ± 0.03	0.22 ± 0.06
MLE 1	26 ± 1.8	6.8×10^5	1.9×10^5	4.0	0.07 ± 0.04	0.09 ± 0.05
MLE 2	26 ± 0.17	6.9×10^5	1.5×10^5	4.1	0.06 ± 0.02	0.07 ± 0.03
Step-feed BNR 2	29 ± 0.18	2.2×10^6	2.9×10^5	14	1.5 ± 0.02	1.7 ± 0.02
Oxidation ditch	19 ± 0.58	3.9×10^5	4.3×10^4	3.4	0.03 ± 0.01	0.03 ± 0.01
Step-feed BNR 3	24 ± 0.78	7.8×10^6	8.6×10^5	57	0.05 ± 0.03	0.06 ± 0.03

*Flow normalized emission factors for centrate are inappropriate since centrate constitutes a miniscule flow rate, while containing up to 30% of the influent TKN load.

From the emission numbers presented in Table 15.3 the following observations can be made:

- On average, N_2O emission fractions varied from 0.01–1.8% or 0.01–3.3%, when normalized to influent TKN load or influent TKN load processed, respectively.
- These emission fractions were on the lower end of the range reported by previous studies, which varied between 0–15% of influent TKN load (Czepiel et al., 1995, Kampschreur et al., 2008b, Kimochi et al., 1998, Sommer et al., 1998, Sümer et al., 1995, Wicht and Beier, 1995).
- The emission fractions are mostly statistically higher (at the $\alpha = 0.05$ confidence level) than currently used values of 0.00035 kg N_2O-N/kg N(non-BNR processes (Czepiel et al., 1995)).

[2]At the first two facilities sampled, both discrete short-term (lasting about 30 min) and continuous (lasting 24 hours) N_2O measurements were conducted in multiple locations or zones. However, subsequently, the 30 min discrete measurements are not reported, since they did not capture the considerable temporal variability observed in each zone. For following campaigns, each location or zone was subjected only to 24 diurnal continuous monitoring.

- Emission values from the separate-stage BNR and oxidation ditch process were statistically lower, (at the $\alpha = 0.05$ confidence level) and the those from the two MLE processes were statistically not dissimilar, (at the $\alpha = 0.05$ confidence level) relative to the current estimates (Czepiel et al., 1995, USEPA, 2009).

The high emissions for the separate centrate treatment process are primarily because centrate streams have disproportionately low flow rates compared to their TKN concentrations.

A high degree of diurnal variability in N_2O emissions was also observed during a day at the full-scale step-feed BNR process (Figure 15.7) This variability could be linked to diurnal variations in influent N-loading as reported by (Ahn et al., 2009).

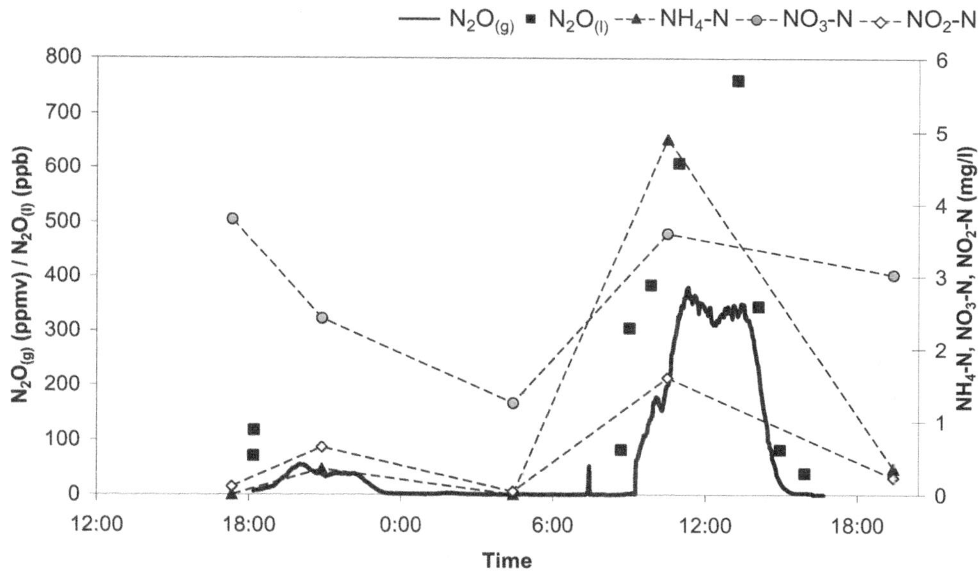

Figure 15.7 Diurnal variability in gaseous N_2O concentrations measured from an aerobic zone of the full-scale step-feed BNR process.

15.3.2 Origin of N_2O emissions

Based on the measurements at the 12 WWTP the emission of N_2O was in general higher from aerated zones than from non-aerated zones as can be seen from Figure 15.8. Therefore the currently held premise that N_2O emissions from WWTPs mostly occur in the anoxic zones (USEPA, 2009) is not accurate.

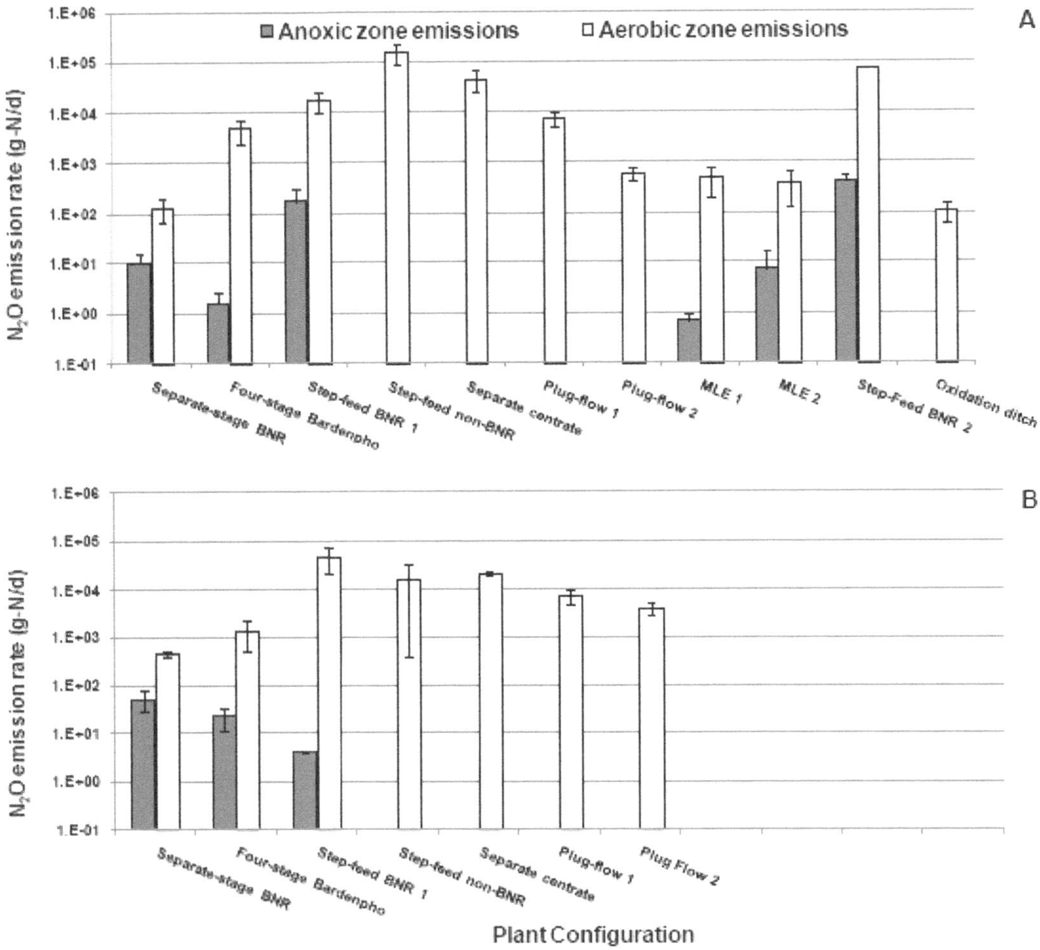

Figure 15.8 N_2O emissions from aerobic and anoxic zones in different WWTPs measured at high (A) and low (B) temperatures. Specific temperatures described in Table 15.3. Step-feed BNR 3 is not included since the emissions from the covered aerobic and anoxic zones could not be distinctly measured.

15.3.3 Process parameters of influence

Aerobic zones

Based on multivariate regression modeling, the factors positively correlated with N_2O emissions from aerobic zones, were NH_4^+-N, NO_2^--N and DO concentrations (isolated effect), and NH_4^+-N and NO_2^--N concentrations (interactive effect) as can be observed from Table 15.4.

Table 15.4 Factors correlating with N_2O emission fluxes (gN_2O-N/day) from aerobic zones.

Variable	Parameter estimate	Standard error	t value	Pr > |t|
Intercept	6.1	0.48	13	0.00
Log(ammonia)	1.0	0.25	4.1	0.00
Log(nitrite)	0.60	0.16	3.7	0.00
Log(DO)	0.59	0.18	3.2	0.00
Log(ammonia)-log(nitrite)	0.18	0.08	2.3	0.03

The outcome of the multivariate regression modeling is also supported by the results from the Bardenpho plant as can be observed from Figure 15.9.

Anoxic zones

Based on multivariate regression modeling, the factors positively correlated with N_2O emissions from anoxic zones, was the DO and NO_2^--N concentration (interactive effect) as can be observed from Table 15.5.

Table 15.5 Factors correlating with N_2O emission fluxes (gN_2O-N/day) from anoxic zones.

Variable	Parameter estimate	Standard error	t value	Pr > |t|
Intercept	-1.2	0.89	-1.3	0.21
Log(DO)-log(nitrite)	0.67	0.25	2.7	0.01

The regression model did not correlate temperature and N_2O fluxes in either aerobic or anoxic zones.

Transition from anoxic to aerobic zones

Besides the parameters correlating to the emission of N_2O another factor of influence could be the transition from anoxic zones to aerobic zones. This was illustrated at different type of plants (processes) as can be observed from Figure 15.10 and Figure 15.11. From Figure 15.10 it can be observed that from the anoxic zone to the aerobic zone the concentration of N_2O in the gas phase increases as well as in the aqueous phase. The same increase in the gas phase was seen in the plug flow reactor (Figure 15.11), but here the liquid-phase N_2O concentrations at both the influent end and middle regions of the passes were statistically similar ($p = 0.26$ and 1.0, respectively for two-tailed t-test comparisons at $\alpha = 0.05$ for the two processes).

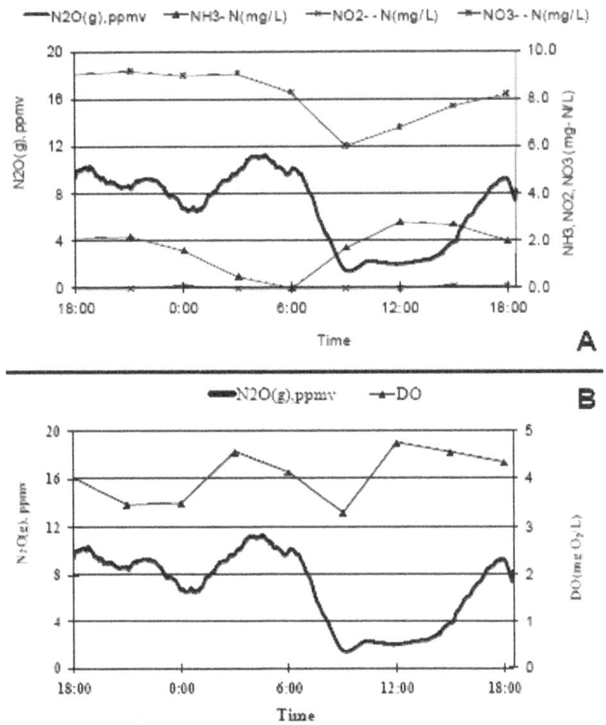

Figure 15.9 Diurnal variability in gaseous N_2O concentrations measured from the first aerobic zone (A-B) of the full-scale four-stage Bardenpho process

Species	Anoxic	Aerobic 1	Aerobic 2
NH_4^+ (mg-N/L)	14	12 ± 5	1.5 ± 0.71
NO_2^- (mg-N/L)	0.00 ± 0.00	0.00 ± 0.00	0.00 ± 0.00
NO_3^- (mg-N/L)	0.85 ± 0.10	2.7 ± 0.35	10 .± 0.21
DO (mg O_2/L)	0.10	2.3	4.2
Aqueous N_2O (μg/L)	55	190	570
Gaseous N_2O (ppmv)	1.5 ± 0.14	16 ± 0.27	23 ± 0.67

Figure 15.10 Spatial profile of gaseous N_2O concentrations and typical activated sludge variables in a Step-feed BNR 1 process showing individual sampling locations. Results are from discrete sampling over a period of 30 minutes at each sampling point. Arrows indicate wastewater flow. Shaded and non-shaded boxes represent non-aerated and aerated zones, respectively. Gaseous N_2O concentrations are expressed as avg. ± sd. of 30 measurements.

Species	Aerobic 1 →	Aerobic 2 →
NH_4^+ (mg-N/L)	10 ± 0.99	6.2 ± 2.2
NO_2^- (mg-N/L)	0.27 ± 0.10	0.22 ± 0.00
NO_3^- (mg-N/L)	1.2 ± 0.42	2.7 ± 1.1
DO (mg O_2/L)	0.80 ± 0.47	1.8 ± 1.0
Aqueous N_2O (μg/L)	56 ± 9.8	62 ± 6.0
Gaseous N_2O (ppmv)	2.9 ± 0.73	51 ± 21

Figure 15.11 Spatial profile of gaseous N_2O concentrations and typical activated sludge variables in a full-scale plug-flow process. Results are from diurnal sampling over a period of 24 hours at each sampling point. Arrows indicate wastewater flow. All zones were aerated.

15.4 NITROUS OXIDE EMISSION THE NETHERLANDS

15.4.1 Emission of N_2O

The N_2O emission that was measured during the four measuring campaigns at the three WWTPs is summarized in Table 15.6.

Table 15.6 Overview N_2O emission at the surveyed WWTPs.

Location	Date	N_2O-N emission (g N_2O-N/h)	NKj influent (kg/d)	Emission[1] (%)
Papendrecht	25-09-08	7.07	n.a.	-
Carrousel	26-09-08	7.69	484	0.038
	27-09-08	7.78	n.a.	-
	28-09-08	6.62	n.a.	-
	29-09-08	6.73	382	0.042
	Average	**7.18**	**433**	**0.040**
Kortenoord	29-05-09	7.4	839	0.021
Carrousel	30-05-09	28.9	965	0.072
	01-06-09	10.7	833	0.031
	02-06-09	24.4	927	0.063

	03-06-09	32.5	n.a.	-
	Average	**20.8**	**891**	**0.048**
Kralingseveer	14-10-08	346	2,348	0.35
BNR	15-10-08	504	2,141	0.57
(October 2008)	16-10-08	294	2,169	0.33
	17-10-08	256	2,642	0.23
	18-10-08	446	2,239	0.48
	19-10-08	436	2,339	0.45
	20-10-08	517	2,295	0.54
	Average	**400**	**2,310**	**0.42**
Kralingseveer	11-02-09	7,900	5,251	3.6
BNR	12-02-09	13,094	5,968	5.3
(February 2009)	13-02-09	11,464	2,998	9.2
	14-02-09	9,733	2,524	9.3
	15-02-09	4,528	2,359	4.6
	16-02-09	8,172	2,331	8.4
	Average	**9,148**	**3,572**	**6.1**
Emission factor NIR[2]				1

[1]The uncertainity in the emission is approximately +/- 5% for Kortenoord and Papendrecht and +/-6% for Kralingseveer.
[2]National Inventory Reports.

From the emission numbers presented in Table 15.6 the following observations can be made:

- The emission found at Papendrecht and Kortenoord were around 20 times lower than the emission factor currently used to estimate the emission of N_2O from WWTPs.
- The emission factor found at Kralingseveer in October was around two times lower than the current used emission factor.
- In February at Kralingseveer the emission was more than six times higher than the current used emission factor.
- The emission of N_2O differ at two WWTPs during one week (Kortenoord and Kralingseveer)
- The emission of N_2O differ between seasons at one WWTP (Kralingseveer)
- The emission of N_2O differ between WWTPs (Papendrecht and Kortenoord versus Kralingseveer)

Besides the observed variations in N_2O emission as mentioned above, also a strong variation was observed during a day at one WWTP. An example of this is presented in Figure 15.11. In Figure 15.11 the variation in N_2O emission is presented as observed during the measuring campaign in October at Kralingseveer.

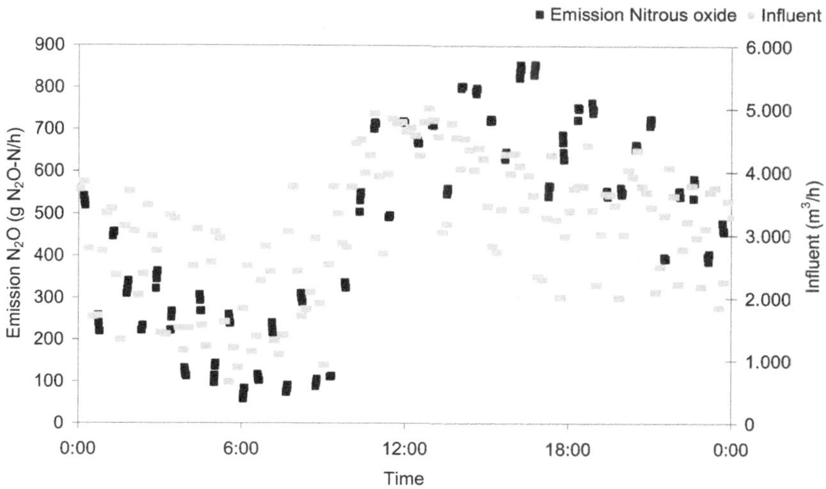

Figure 15.11 N_2O Emission and influent flow at 18-10-2008 at WWTP Kralingseveer.

From Figure 15.11 it can be observed that the emission of N_2O varies with the influent flow. This suggests that the emission of N_2O is related to the variation in sludge load. However, a linear relation between sludge load and N_2O emission could not be found at Kralingseveer, but this relation was found in Kortenoord as can be seen from Figure 15.12.

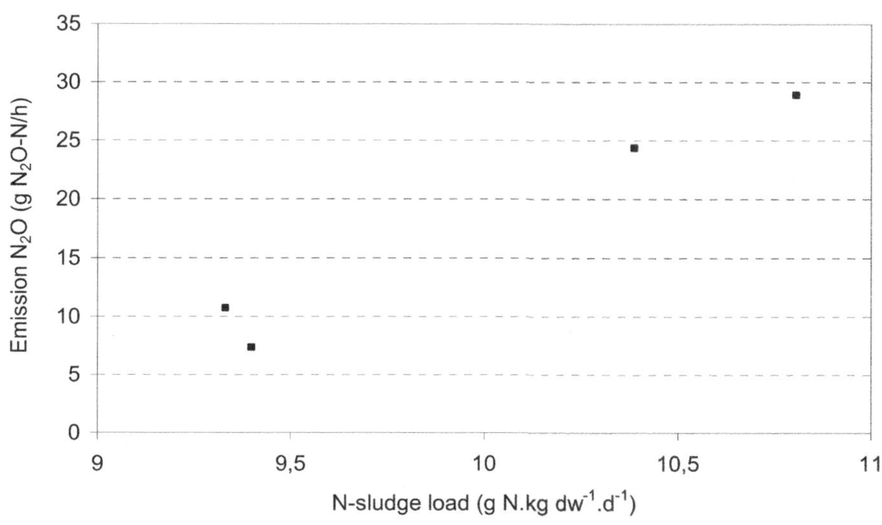

Figure 15.12 The emission of N_2O as function of the N-sludge load as observed during the measuring campaign in Kortenoord.

15.4.2 Origins of N_2O emission

One of the main objectives of the research performed in the Netherlands was to investigate to which extent WWTPs emitted N_2O and what the dynamics of this emission was. To study this, the emission of N_2O was measured in the total off gas of WWTPs that are completely covered. The total off gas originated from tanks in which both nitrification and denitrification processes took place. For this reason it was not possible to distinguish which process nitrification or denitrification was responsible for the emission of N_2O.

15.4.3 Process parameters of influence

Based on the measuring campaigns differences in N_2O emissions were found between WWTPs and differences where found during day or in between seasons at one WWTP.

Possible process parameters that could explain these differences are summarized in Table 15.7.

Table 15.7 Overview process parameters that could explain the found variations in N_2O emission.

Variation in N_2O emission	Possible process parameters of influence
During a week at one WWTP (p.e. Kortenoord)	Nitrogen sludge load
During seasons at one WWTP (Kralingseveer)	Temperature
	Rainfall
	Aeration intensity
	Nitrite
	Nitrogen sludge load
Between WWTPs Papendrecht/Kortenoord vs Kralingseveer	BOD / N ratio
	Nitrogen sludge load
	Type of process

The difference in N_2O emission at WWTP Kortenoord could possibly explained by the variation in nitrogen sludge loading (see Figure 15.12).

At the WWTP Kralingseveer the emission of N_2O was measured in October and February to study the effect of the temperature on the emission. In October the average temperature of the wastewater was 18.1°C, in February the temperature was 9.6°C. However the effect of only the temperature could not studied, because at the start of the measuring campaign in February the WWTP suffered from heavy rainfall. At that time the WWTP had to treat wastewater flows of almost 13,000 m^3/h, while the dry weather flow is normally ~5,000 m^3/h. As a result of the heavy rainfall the nitrogen sludge load increased and the aeration intensity was increased to remove at least all the ammonium. Because different events (low temperature together with heavy rainfall) coincide at the same time during the measuring campaign in February it was not possible to point out one cause for the very high N_2O emission in February compared to the emission in October. What was found in February was a correlation between the N_2O emission and the nitrate concentration as can be observed from Figure 15.13. The found correlation between the nitrate concentration and the N_2O emission suggest that also nitrite was present. Nitrite is not a parameter that is measured at a WWTP, but based on a simple analysis during the measuring campaign in February the presence of nitrite was proven and concentrations up to ~10 mg/l were found.

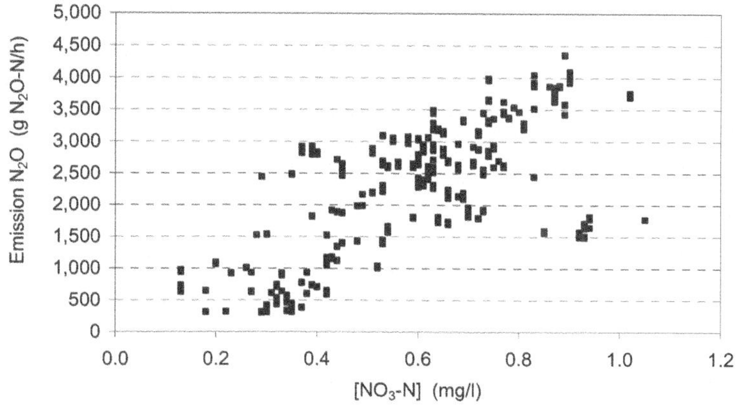

Figure 15.13 The emission of N_2O as function of the nitrate concentration as observed in the first aeration tank at WWTP Kralingseveer (February 2009).

The average N_2O emission found at WWTP Papendrecht and Kortenoord were comparable, but the emission at Kralingseveer (October) was almost ten times higher. The following differences between Papendrecht and Kortenoord on one side and Kralingseveer on the other side could possibly explain the difference in N_2O emission:

- Absence of primary clarifier at Papendrecht and Kortenoord and the presence of a primary clarifier at Kralingseveer. This resulted in a difference in the BOD/N ratio. During the measuring campaign in Papendrecht the BOD/N ratio was 2.9, in Kortenoord this ratio was 2.5 and in Kralingseveer this ratio was 1.7 in October, but was 2.7 in February.
- A lower nitrogen sludge load in Papendrecht and Kortenoord compared to Kralingseveer. During the measuring campaign in Papendrecht the nitrogen sludge load was 0.012 $kgN \cdot kg\ dw^{-1} \cdot d^{-1}$, in Kortenoord this was 0.010 $kgN \cdot kg\ dw^{-1} \cdot d^{-1}$, and in Kralingseveer this was 0.020 $kgN \cdot kg\ dw^{-1} \cdot d^{-1}$ in October, and in a February a nitrogen sludge load of 0.053 $kgN \cdot kg\ dw^{-1} \cdot d^{-1}$ was observed.
- Process configuration. In Papendrecht and Kortenoord the removal of nitrogen occurs in a carrousel, while in Kralingseveer nitrogen is removed in first a plug flow reactor followed by two carrousels.

15.5 METHANE EMISSION AUSTRALIA

15.5.1 Liquid phase data

Rising main UC09

Rising main UC09 is a single line receiving domestic wastewater. It has a length of 1080 m and a pipe diameter of 150 mm (Area to Volume ratio: $A/V = 2/0.075 = 26.7\ m^{-1}$). The average daily flow was approximately 200 m^3/d at the time of the study, giving rising to an average HRT of 2.5 hr. However, HRT varied between 1.5 to 6.0 hours caused by diurnal variations of the wastewater flow. The feed pump was operated intermittently with an interval ranging between below 30 minutes to above 2 hours. Each pumping event lasted for 1-3 minutes depending on the time of the day. The average flow velocity during a pumping event was estimated to be 1.55 m/s. The examination of a removable section of the pipe revealed that no sediments were present in this pipe.

Figure 15.14 shows the dissolved CH_4 concentrations measured at both the pumping station and 828 m downstream from the pumping station during daytime (10:00 a.m.–6:00 p.m.) over a time span of two years.

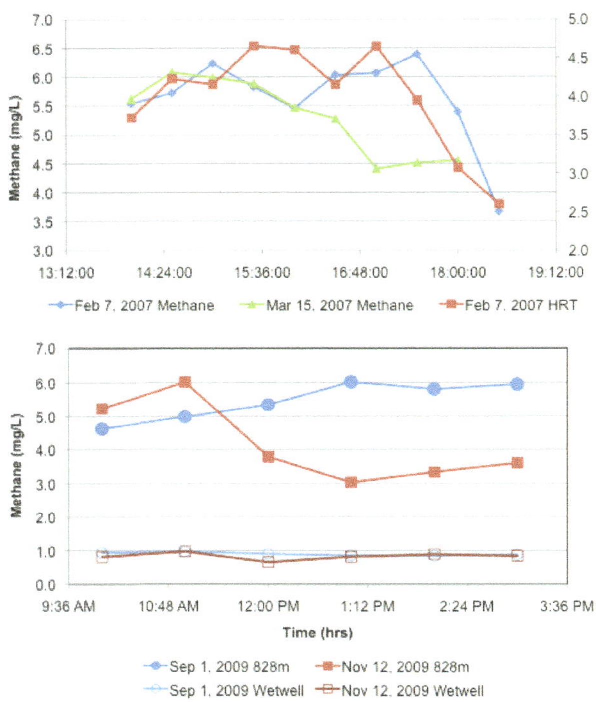

Figure 15.14 Methane data measured at UC09. Part of the data was previously published in Guisasola et al. (2008). The wastewater temperature was 27.7°C, 26.9°C and 24.6°C on 07/02/2007, 15/03/2007 and 01/09/2009, respectively.

In one case, the HRT of the wastewater samples taken at 828 m was calculated and also shown in the figure. The HRT of a wastewater sample, defined as the time of this sample spent in the rising main at the time of sampling, was calculated from the operational data of the pump feeding the given rising main (Sharma et al., 2008). The volume of wastewater pumped into the pipe during each pump run was calculated from the physical dimensions of the wet well and the measured stop/start water levels. Based on this estimated wastewater volume and the dimensions of the pipe, the number of pump runs required for a wastewater "slug" to travel from the pipe inlet to the sampling location was calculated. The HRT of the wastewater sample collected was then calculated from the on-line pump operational data recorded by the on-line supervisory control and data acquisition (SCADA) system.

In addition to CH_4 concentrations, several other physico-chemical parameters of the samples collected such as temperature, total and soluble COD, volatile fatty acids (VFAs), inorganic sulphur species (sulphate, sulphide, thiosulphate and sulphite) and pH were also measured.

Methane concentration at the pumping station was always approximately 1 mg/l. In comparison, CH_4 concentration at 828m was between 3–6.5 mg/l with an average of approximately 5.0 mg/l, indicating an average production of 4.0 mg/l. The average CH_4 production rate was 1.3 $mg \cdot l^{-1} \cdot hr^{-1}$, or 1.2 $g \cdot m^{-2} \cdot d^{-1}$ by taking the A/V ratio of 26.7 m^{-1} into consideration.

Rising main C016

A second sampling site is CO16, which were sampled at four locations, namely the pumping station and 500m, 1100m and 1900m downstream of the pumping station (Figure 13). Similar to UC09, the CO16 rising main also receives domestic wastewater. It has an internal pipe diameter of 300 mm, giving an A/V ratio of 13.3 m^{-1}. The average daily wastewater flow was approximately 700 m^3. Similar to UC09, the pumping station was operated intermittently with 30–40 pumping events per day, each lasting for 4–6 min in duration. During each pumping event, the calculated flow velocity was approximately 0.90–1.01 m/s.

In one sampling campaign, samples from the four locations were collected hourly between 5:00 a.m. and 8:00 a.m. to cover a wide range of HRT (0 to 8.7 hr), as the samples captured included wastewater discharged over night and hence with a relatively long retention time in the sewer line, and also relatively fresh sewage discharged in the morning. The CH_4 data are plotted against the sampling locations in Figure 15.15 (Foley et al., 2009). More detailed data can be found in Foley et al. (2009).

Similar to UC09, CH_4 concentration in the pumping station was between 1-2 mg/l. The concentration increased further downstream, reaching approximately 9.0 mg/l. Methane concentration varied considerably with time at the 500m and 1100m locations between 5am and 8am, likely due to the variation of HRT of the wastewater samples collected. The wastewater HRT at 500m reduced from 5.4 hr at 5:00 a.m. to just below 1 hr at 7–8:00 a.m. Similar, the wastewater HRT reduced from 7.3 hr at 4:00–5:00 a.m. to 1.5 hr at 7:00–8:00 a.m. By taking all results together, the average CH_4 production rate was calculated as 0.7 $mg \cdot l^{-1} .hr^{-1}$ or 1.3 $g \cdot m^{-2} \cdot d^{-1}$. The latter is very similar to the value observed at UC09.

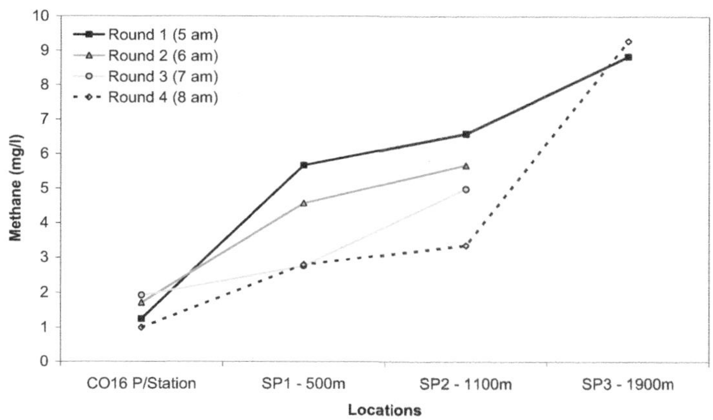

Figure 15.15 Methane data measured at four locations along CO16 (Foley et al., 2009). The average wastewater temperature was 23.5°C.

In another sampling campaign, the 1100m location was sampled half-hourly during 4:15 a.m. and 10:00 a.m. The results are presented in Figure 15.16.

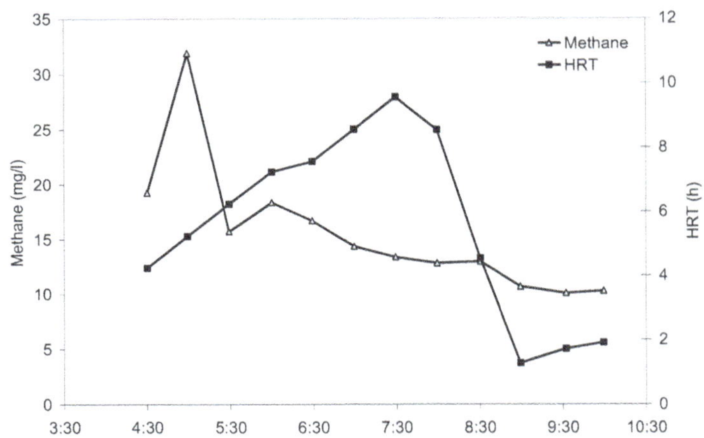

Figure 15.16 Methane data measured at CO16 at 1100m in another campaign. The average wastewater temperature was 22.5 °C.

The CH_4 concentrations measured this round were approximately three times those presented in Figure 15.15. This difference cannot be fully explained by the differences in HRT (4–9.5 hr in this case in comparison with 1.5–7.3 hr in Foley et al., 2009). The average CH_4 production rate was calculated as 3.1 $mg \cdot l^{-1} \cdot hr^{-1}$ or 5.5 $g \cdot m^{-2} \cdot d^{-1}$, three times higher than the values obtained in the previous round or for UC09. The average wastewater temperature was 22.5 °C, which was similar to that in the previous round. The reason for the much higher CH_4 production in this round is not known. This could be related to the COD (Chemical Oxygen Demand) concentration, which unfortunately was not measured during the measurement campaigns

Laboratory-scale sewer systems

To investigate CH_4 production in sewers under controlled conditions, reactor-based laboratory scale systems were set up at The University of Queensland (Figure 15.17).

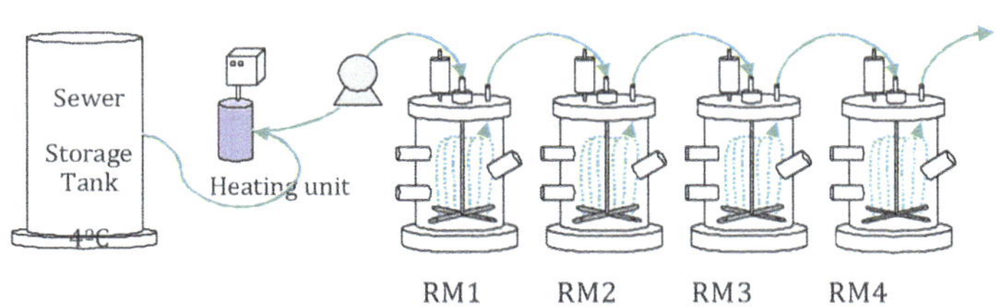

Figure 15.17 A laboratory-scale sewer system at The University of Queensland.

Each system consisted of four air-tight reactors in series (named RM1 to RM4), each with a volume of 0.75 L. The diameter of the reactors was 80 mm and the height was 150 mm. The A/V ratio was calculated to be 56.7 m^{-1} by considering the biofilm areas both on the wall and on the top of the reactor (there was no visible biofilm growth on the bottom of the reactor due to the use of a magnetic stirrer, see below). Biofilm carriers were placed in each reactor to enable the collection of intact biofilms for micro-scale analysis. This increased the biofilm surface area by 25%, and the effective A/V ratio was thus increased to 70.8 m^{-1}. The system was intermittently fed with real sewage that had been screened and settled using a peristaltic pump following a typical operating pattern observed at UC09 in 2003. The feed pattern consisted of sixteen pumping events per day, each delivering 0.75 l (equivalent to the volume of one reactor) wastewater into the system. The average HRT in the system was 6 hrs with minimum and maximum HRT being 2 and 10 hrs, respectively. The system was operated under quiescent conditions except during pumping events, which resulted in turbulence in the reactors due to flow. The reactors were also stirred via magnetic stirrers during pumping events to minimize the accumulation of sediments.

Fresh sewage was collected weekly at the Robertson Park Pump Station, Indooroopilly (Brisbane), which primarily collects domestic wastewater from the local area, and thus contains a low amount of sulphide ($\leq$ 1 mg S/l) and CH_4 (~1 mg/l). The fresh sewage contains on average 15 mg SO_4-S/l and 200 mg COD/l of soluble COD, of which the Volatile Fatty Acid (VFA) content was approximately 50 mg COD/l as acetate and 15 mg COD/l as propionate. Sewage was stored at 4°C to minimize biological transformations and warmed to 21°C through a heat exchanger and water bath prior to entering the system. The sewage temperature remained stable as the reactor system was located in a temperature controlled lab (21 ± 1°C).

After several months of operation, anaerobic biofilm was fully developed in the laboratory system, leading to pseudo-steady state conditions. The system was monitored with routine sampling of wastewater from all reactors. Figure 15.18A shows typical CH_4 profiles in all reactors during daytime.

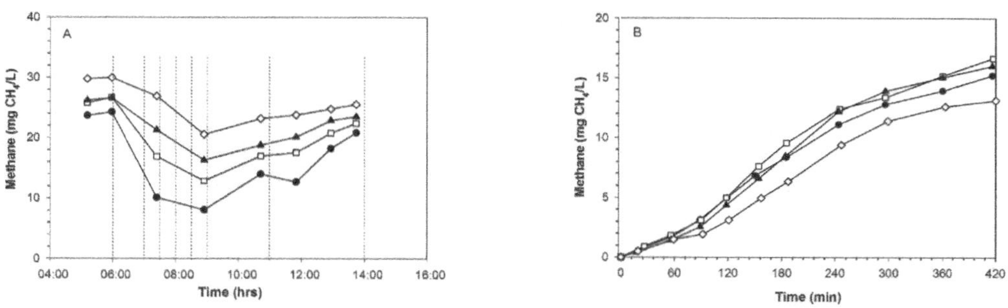

Figure 15.18 A. Spatial and temporal variations of CH_4 concentrations in the laboratory- scale sewer systems; B. Methane profiles obtained in batch tests with the four reactors. RM1 (•), RM2 (□), RM3 (▲) and RM4 (◊). Time 0 is 0am. The vertical dashed lines in A corresponded to pumping events. Wastewater temperature during the test was approximately 21°C.

The results confirmed that a substantial amount of CH_4 can be produced in anaerobic sewer systems. The measured CH_4 concentrations were between 8 - 30 mg/l, generally higher than the values measured in real sewers. This was likely due to the higher A/V ratio of the lab-system (70.8 m^{-1} vs. 13.3 m^{-1} of CO16 and 26.7 m^{-1} of UC09). As will be further discussed later, however, the CH_4 production rate per area of sewer biofilm (calculated to be 1.0–1.2 g $CH_4 \cdot m^{-2} \cdot d^{-1}$) is very similar to the values determined for UC09 and CO16.

Batch tests were performed with all the four reactors to assess the methanogenic activities of the biofilms. The continuous operation of the system was stopped and the experimental reactors were temporarily isolated from the system. At the start of each batch test, the selected reactor was carefully emptied and drained to remove any sediment, and re-filled with fresh sewage, taking care to minimize oxygen entrainment during filling. The reactor was magnetically stirred during the batch experiments. As an average, two batch tests per month were conducted in each reactor.

Each batch test typically lasted around eight hours. Methane, VFAs, COD and soluble sulphur species (sulphate, sulphite, thiosulphate and sulphide) were measured regularly. A sample frequency of around 30 min was maintained for at least the first three hours of the test, while the last 5 hours were usually sampled on an hourly basis. Figure 15.18B shows the typical CH_4 profiles obtained during the batch tests. Methane production was approximately linear in all reactors in the first 5 hours. The slowing-down after 5 hours was likely due to the decrease of easily fermentable COD (Guisasola et al., 2008). The relatively slow rates in the first 1.5 hours were not expected, which could be due to the contact of anaerobic biofilm with oxygen during the draining process. Very similar CH_4 production rates were observed for the first three reactors. The fourth reactor had a slightly lower rate. The average CH_4 production rate was 2.9–3.7 $mg{\cdot}l^{-1}.hr^{-1}$ (determined using the linear part of the curves presented in Figure 15.18B, i.e. between 90–240 minutes), or equivalently 1.0–1.2 $g{\cdot}m^{-1}{\cdot}d^{-1}$. These values are very similar to those observed at UC09 and CO16.

15.5.2 Gas phase data

While it is a routine process for sewer workers to measure %LEL levels in confined spaces, CH_4 concentration in sewer air has not been widely reported in literature. In the last few years, South East Water Limited (SEWL) investigated CH_4 production in one of its sewer mains (Ibrahim, 2010). The main is 1.8 km in length. The pipe material is concrete and diameter starts at 150mm and goes to 525 mm. The main receives wastewater from a number of industrial dischargers and a small amount of domestic sewage towards the end of main. South East Water Ltd. detected high %LEL levels and low O_2 readings during scheduled cleaning of this main in January 2006, which triggered a closer monitoring of the %LEL levels and CH_4 concentrations in the manholes along the sewer line. Figure 15.20 shows the %LEL levels at various manholes over a period of 1.5 years.

Very high %LEL levels were repeatedly detected in this sewer, particularly at downstream locations. In several occasions, %LEL was above 100%, which implied a risk of explosion. Methane was confirmed to be responsible for the high %LEL readings through bag sampling and gas composition analysis with GC. In the two samples taken from Manhole M6 at 2am and 2pm, CH_4 concentrations were measured as 1.7% (34% LEL) and 0.6% (12% LEL), respectively.

The fact that %LEL was relatively low at upstream locations (M11 and M10B) suggests that CH_4 emitted at downstream locations was formed in this gravity section rather than as result of CH_4 formation in a pressurized section located further upstream, as otherwise high emissions would have been observed at M11 and M10B as well.

It should be noted that this system may not be representative to general sewer networks due to the industrial origin of the wastewater transported. Indeed, as will be further discussed, SEWL was able to identify the cause for the high-level CH_4 formation, which is linked to a particular food processing wastewater that contained a high-level of biodegradable COD. This will be further discussed later.

However, high-levels of CH_4 can also be present in sewer networks that primarily transport domestic wastewater. Indeed, high concentrations of CH_4 would be expected at discharging points of rising mains. Figure 15.20 shows that the dynamic CH_4 and oxygen profiles measured at the discharging well of CO16, where high-levels of dissolved CH_4 were detected (referring to section 6.5.1).

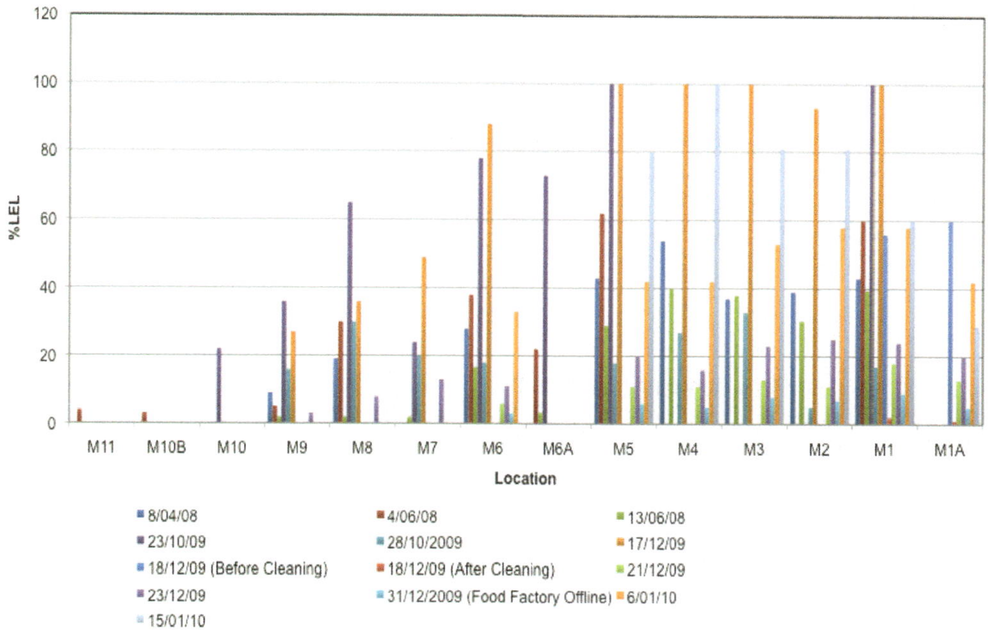

Figure 15.20 Gas phase %LEL in manholes along a main receiving predominantly industrial wastewaters (Ibrahim, 2010).

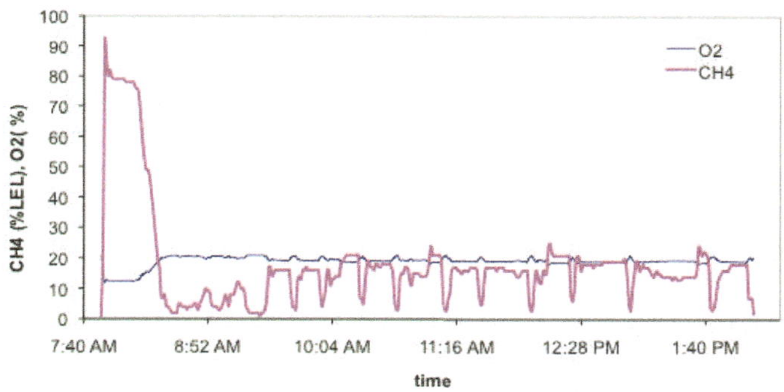

Figure 15.21 Gas phase CH_4 %LEL and oxygen concentration at a rising main discharge point.

In the early morning, a %LEL level of 80–90% was measured, which was accompanied by a low oxygen level of 12–13%. The very high level of CH_4 can be explained by the discharge of septic wastewater that had stayed in the rising main for many hours over night. The %LEL levels varied between 2–22% during the remaining time of the monitoring period. The dynamics was likely caused by the intermittent discharge of CH_4-containing wastewater from the rising main. Unfortunately, the monitoring trial lasted for a few hours only. More on-line data of this nature should be collected in the future.

15.5.3 Modelling

An empirical model

Both field data and simulation studies revealed that, for a given wastewater composition, the key factors determining CH_4 formation in a rising main sewer are the HRT of wastewater, and the A/V ratio of the pipe. Based on this observation, Foley et al. (2009) proposed the following empirical equation for estimating CH_4 production in a rising main pipe:

$$C_{CH4,end} = C_{CH4,0} + r_{CH4,A} * \frac{A}{V} * HRT \qquad (1)$$

where $C_{CH4,end}$ and $C_{CH4,0}$ are CH_4 concentrations (mg/l or g/m^3) at the end and at the beginning of the pipe, respectively; A/V = 2/r is the A/V ratio (m^{-1}) of the pipe with r (m) being the pipe radius; HRT is the hydraulic retention time (hr) of wastewater in the pipe; $r_{CH4,A}$ is the CH_4 production rate of sewer biofilms (g/m^2).

Field measurements carried out in Australia so far have consistently shown that $C_{CH4,0}$ has a value of 1-2 g/m^3 in wet wells collecting domestic sewage.

With the field data measured, $r_{CH4,A}$ was empirically determined to be 0.0524 g·m^{-2}·hr^{-1} in Foley et al. (2009), which was shown to agree well with the Guisasola et al. (2008) model (Foley et al., 2009). However, the value of this parameter is expected to be largely affected by, among other factors, wastewater composition and temperature and may need calibration in each case.

This simple predictive model provides a means for water authorities to estimate CH_4 formation from their pressurized sewerage systems.

Limitations of current models

The modelling work to date has focused on rising main sewers. In gravity sewers, both CH_4 formation and oxidation are expected to occur due to the presence of surface aeration of wastewater. However, to date there has been no study on CH_4 transformation in gravity sewers. In particular, the contribution of sewer sediments to CH_4 formation is currently unknown. This should be addressed in future research.

While there has been no study reported to date dedicated to the gas-liquid mass transfer of CH_4 in sewers, its modelling should be relatively straightforward given the large amount of work undertaken to model the transfer of oxygen and hydrogen sulphide in sewer systems (Hvitved-Jacobsen, 2002).

15.5.4 Impact of trade waste

Many types of trade wastes contain a high-level of easily biodegradable COD, which may increase CH_4 production in sewers. Indeed a major contributor to the high-levels of CH_4 presented in Figure 15.20 was confirmed to be the wastewater discharged by a food factory. The trade wastewater contained total COD (TCOD) and soluble COD (SOCD) at 26,500 and 30,000 mg/l, respectively, elevating the TCOD and SCOD concentrations from 1,750 mg/l and 1,050 mg/l before the trade waste discharge to 14,500 mg/l and 12,000 mg/l, respectively, after the discharge. Methane generation testing was conducted to determine if this particular trade waste was responsible for the high-level of CH4 measured. Testing was performed with mini-digesters to measure the overnight change in gas pressure in four samples. Samples were taken from the trade waste, upstream and downstream of the trade waste discharge point, as well as at the end of the sewer line. The samples were left overnight at 21°C and the gas pressure measured approximately after 18 hours. The upstream sample showed the lowest reading in pressure (see Table 15.8). A dramatic increase CH_4 production by the downstream samples, along with the high CH_4 production potential of the raw trade waste, suggests that the trade waste was likely a major source of CH_4. This was independently confirmed by the fact that CH_4 concentration in sewer air decreased dramatically when the factory was closed during a holiday period.

Table 15.8 Methane production during 18 hr anaerobic incubation of wastewater samples collected at various locations (Ibrahim, 2010).

Wastewater sample collection location	Final pressure (mm Hg)	Soluble COD (mg/l)	Total COD (mg/l)
Upstream of the trade waste discharge point	22	1,050	1,750
Downstream of the trade waste discharge point	72	12,000	14,500
End of the sewer line	85	9,400	11,000
Trade waste	120	26,500	31,000

An in-depth study of the impact of brewery wastewater discharge on sulphide and CH_4 production in a sewer was conducted in Sudarjanto et al. (2010a) using laboratory sewer systems. Experiments were carried out on two laboratory scale sewer reactors, one used as an experimental and the other as a control. Each reactor had a volume of 1 l, and was fed with wastewater four times a day. During each pumping event, 1 l of wastewater was pumped into each reactor, resulting in a hydraulic retention time (HRT) of 6 hours.

As summarized in Table 15.9, both reactors were fed with domestic sewage for six months to enable the development of stable anaerobic sewer biofilms (baseline period). While the feed to the control reactor remained unchanged, the experimental reactor was fed with a mixture of brewery and domestic wastewater at two different proportions at later stages (10% v/v for three months followed by 25% v/v for another 1.5 months). The composition of the sewage and brewery wastewater is shown in Table 15.10. Batch tests were conducted to measure the CH_4 production rates of the two reactors during the three phases.

Table 15.9 Details of the operational periods (Sudarjanto et al., 2010a)

Phase	Duration	Composition of feed	
		Experimental reactor	Control reactor
1	6 months	100% domestic wastewater	100% domestic wastewater
2	3 months	90% domestic + 10% brewery wastewater	100% domestic wastewater
3	1.5 months	75% domestic + 25% brewery wastewater	100% domestic wastewater

Table 15.10 Composition of sewage and brewery wastewater used in the study (Sudarjanto et al., 2010a)

Parameter	Unit	Composition of feed	
		Raw sewage	Brewery wastewater
TCOD	mg COD/L	270-485	5,000-7,000
SCOD	mg COD/L	170-310	3,000-6,000
VFA	mg COD/L	50–120	250-500
pH	-	7.2 ± 0.2	4.0 ± 0.6

As shown in Figure 15.22, 10% v/v discharge of brewery wastewater increased the CH_4 production rate of the sewer biofilm by 30%.

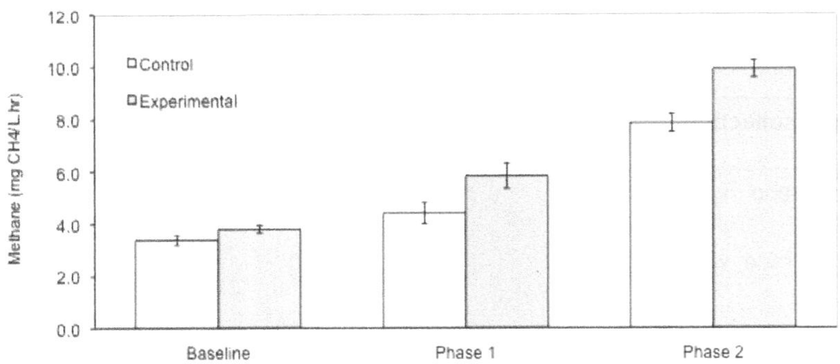

Figure 15.22 Methane production rates by the anaerobic sewer biofilms in the control and experimental reactors during the three phases of the study (Sudarjanto et al., 2010).

When the brewery wastewater fraction was increased to 25% v/v, the CH_4 production rate in the experimental reactor remained to be 30% higher than that in the control reactor. The results confirmed that the discharge of brewery wastewater into sewers significantly increase CH_4 production by sewer biofilms. It is also interesting to observe that the ratio of CH_4 production rates between the two reactors did not increase when the brewery wastewater fraction was increased from 10% to 25%, with reasons yet to be identified. It may be that CH_4 production does not increase further once a critical SCOD is reached in the wastewater. This is understandable considering the fact that biological reactions in sewers are primarily catalyzed by sewer biofilms (including sediments) (Gutierrez et al., 2009), and therefore the reaction rate may be limited by the biofilm surface area. A further contributor could be the low pH in the experimental reactor during brewery wastewater addition.

The average pH in the control reactor was 7.0, which was also the pH in the experimental reactor in the baseline period. With 10% and 25% addition of the brewery wastewater, pH in the experimental reactor decreased to 6.6 and 6.0, respectively. The low pH could have partially reduced the activities of methanogens.

It should be noted that different types of trade wastes may have different effects on CH_4 formation in sewers. In a recent study, Sudarjanto et al. (2010b) found that that the discharge of a dairy wastewater, which also contains high-levels of SCOD and TCOD, at a ratio of 10%v/v did not cause any increase in CH_4 production in a laboratory sewer system. Indeed, both sulphide and CH_4 production was observed to decrease slightly. The results suggest that each trade waste should be assessed separately. Nevertheless, it is conceivable that trade wastes that contain high-levels of carbohydrates will likely enhance CH_4 formation in sewers.

15.5.5 Mitigation

The mitigation of CH_4 emission from sewers is important for the water industry to minimize the carbon footprint of its operation. This would be best achieved through reducing CH_4 production by suppressing the growth of methanogens in sewer biofilms. The mineralization of CH_4 after its production is an alternative option, but may be difficult to achieve in the sewer environment. The biological oxidation of CH_4 is a slow process in general.

Chemical dosage to sewage is commonly used for the control of sulphide formation in sewers, or its transfer from wastewater to sewer air. The effect of several of these chemicals including nitrite, iron salts and magnesium hydroxide on CH_4 formation by sewer biofilms has recently been investigated (Mohanakrishnan et al., 2008; Gutierrez et al., 2009; Zhang et al., 2009; Jiang et al., 2010a). These chemicals substantially reduce or even completely suppress methanogenic activities. Once inhibited or suppressed, it takes weeks or even months for methanogens to fully recover (Jiang et al., 2010a). These results indicate that CH_4 emission can be mitigated in conjunction with sulphide emission. Both CH_4 and sulphide control should be considered when optimizing the use of these chemicals. Future research should also focus on the development of more cost-effective mitigation strategies, perhaps dedicated to CH_4 control.

Elevation of pH

Elevation of pH through the addition of magnesium hydroxide ($Mg(OH)_2$) is commonly used as a means for reducing H_2S emission from wastewater to sewer air. When pH is lifted from a neutral level, which is typical in sewage, to 8.5–9.0, the molecular hydrogen sulphide (H_2S) fraction of total dissolved sulphide is reduced, and as a result its transfer from the liquid to the gas phase is reduced.

Gutierrez et al. (2009) investigated the impact of pH elevation on the microbial activities of anaerobic sewer biofilms, related to sulphide and CH_4 production as well as fermentation.

Experiments were carried out on laboratory scale sewer systems consisting of both an experimental and a control reactor. Each reactor, with a volume of 1 l, was fed with domestic wastewater every six hours.

During each pumping event, 1 l of wastewater was pumped into each reactor, resulting in a hydraulic retention time (HRT) of 6 hours. The 12-month experimental study was divided into four phases with conditions summarized in Table 15.11.

Table 15.11 Experimental conditions applied during the entire course of the pH study (Gutierrez et al., 2009)

Operational phases	Length (days)	Control Reactor	Experimental Reactor
1	0–40	No pH control 7.6±0.1	No pH control 7.6±0.1
2	51–110	No pH control 7.6±0.1	pH adjusted to 8.6±0.1 with 88 ml 0.05M NaOH in each pumping event
3	111–170	No pH control 7.6±0.1	pH adjusted to 9.0±0.1 with 120 ml 0.05M NaOH in each pumping event
4	171–332	No pH control 7.6±0.1	No pH control 7.6±0.1

Compared to those in a control reactor without pH control (pH 7.6±0.1), the SRB activity was reduced by 30% and 50%, respectively, at pH 8.6 and pH 9.0. The results showed that pH elevation not only reduced the H_2S transfer but also its production by sewer biofilms. When normal pH was resumed, it took approximately two months for the SRB activity to fully recover. More details can be found in Gutierrez et al. (2009).

Methanogenic activities developed in the control reactor in three months after the reactor start-up, while no significant methanogenic activities were detected in the experimental reactor until normal pH was resumed (Figure 15.22). The results suggest that elevated pH at 8.6–9.0 suppressed the growth of methanogens.

These experimental findings are significant for the optimal use of alkali addition to sewers for the control of H_2S and CH_4 emissions. A model-based study performed in Gutierrez et al. (2009) showed that, by adding the alkali at the beginning rather than towards the end of a rising main, substantial savings in chemicals can be achieved while achieving the same level of sulphide emission control, and complete CH_4 emission control.

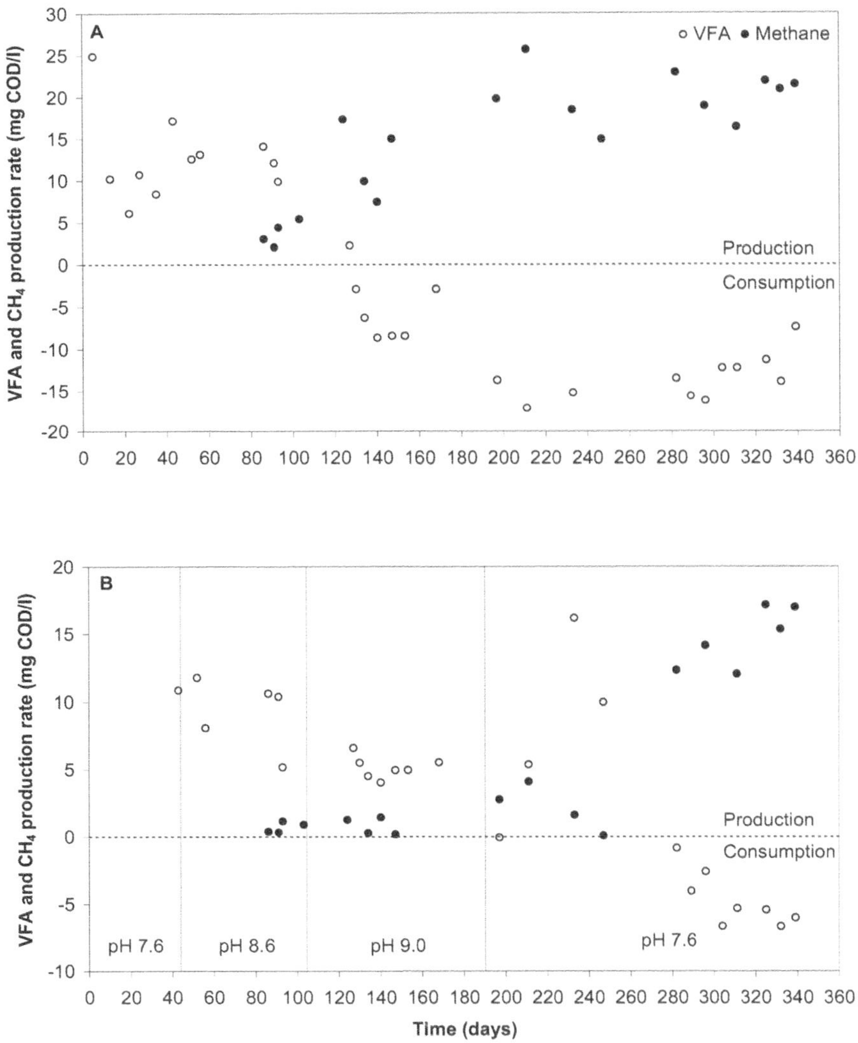

Figure 15.23 Evolution of the VFA (○) and CH_4 production rates (•) in the control (A) and experimental (B) reactors. Note that CH_4 production rates have a unit of mgCOD·l^{-1}·hr^{-1} in this figure, rather than mgCH_4·l^{-1}·hr^{-1}, for direct comparison with the VFA production/consumption rates. 1 mgCH_4 = 4 mg COD.

Elevated pH conditions also reduced the activity of fermentative bacteria (FB) in the reactors. Prior to the development of the methanogenic activity, a stable VFA production rate of 11.5±1.2 mg VFA-COD·l^{-1}·hr^{-1} was observed in the control and experimental reactors.

The FB activity decreased to 5.1±0.4 mg VFA-COD·l^{-1}·hr^{-1} in the experimental reactor under pH 9.0 (with negligible CH_4 formation). This indicates that the fermentation activity was reduced by 54% at pH 9.0 in comparison to pH 7.6.

Nitrite addition

Mohanakrishnan et al. (2008) and Jiang et al. (2010a) investigated the impact of nitrite addition on the sulphate-reducing and methanogenic activities of anaerobic sewer biofilms. It was found that intermittent addition of nitrite could suppress CH_4 production for weeks to months.

The laboratory systems used in Jiang et al. (2010a) are shown in Figure 15.24, which consisted of four air-tight reactors, namely R1 to R4, each with a volume of 0.75 l. The reactors were fed with sewage through a peristaltic pump every 6 hours, a typical sewage hydraulic retention time in sewers (Hvitved-Jacobsen, 2002).

Every feed pumping event lasted for 2 minutes, delivering one reactor volume (0.75 l) of sewage into each reactor.

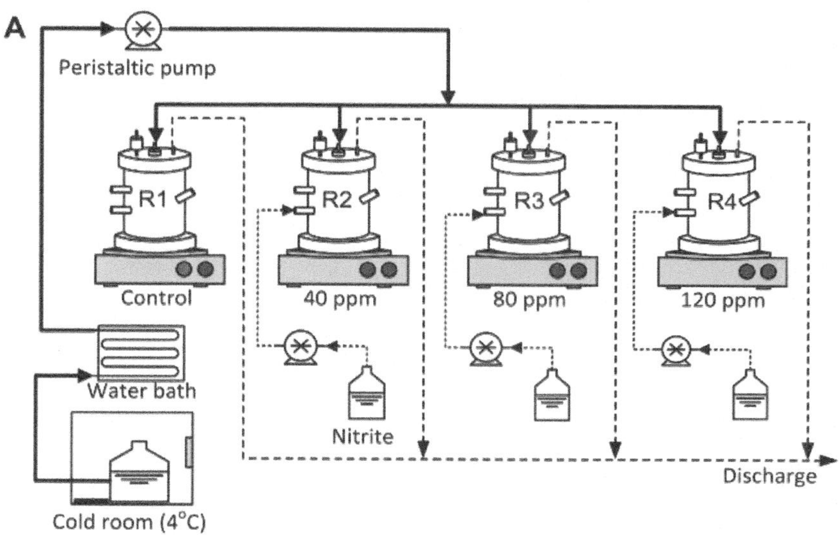

Figure 15.24 Schematic of the laboratory-scale rising main sewer reactors. R1 did not receive nitrite, and served as a control. R2, R3 and R4 received nitrite at 40, 80 and 120 mgN per L of wastewater fed, for a period of 24 days, which was followed by a 2.5 month recovery period.

The experiments were conducted in three consecutive phases, namely the stabilization, dosing, and recovery phases. Reactors were operated without nitrite dosing to achieve similar sulphide and CH_4 production activities during the stabilization phase. During the dosing phase, R2-R4 received nitrite, while R1 was used as the control reactor (no nitrite dosage). Reactors R2-R4 were injected with nitrite to reach concentrations of 40, 80, and 120 mg-N/l of wastewater, respectively.

After the 24-day dosing phase, nitrite dosing to R2 - R4 was stopped and the reactors were allowed to recover for two months (Recovery phase).

Methane production in dosed reactors was reduced to negligible levels within 2 days after the initiation of nitrite dosage (Figure 15.25). Note that the methanogenic activity was measured in the absence of nitrite through batch tests using fresh sewage without nitrite addition in all cases. The different levels of nitrite addition did not cause any difference in inhibition. Complete methanogenic inhibition was achieved with the lowest nitrite concentration used in the experiment, i.e. 40 mg-N/l. The recovery of CH_4 production proceeded at very similar rates in all cases. The recovery process was almost linear during the whole recovery phase. Only less than 60% of recovery was achieved after two-month of recovery. Suppression of sulphide production was also achieved with nitrite dosage at 80 mg-N/l and 120 mg-N/l. However, the recovery of sulphide production was considerably faster.

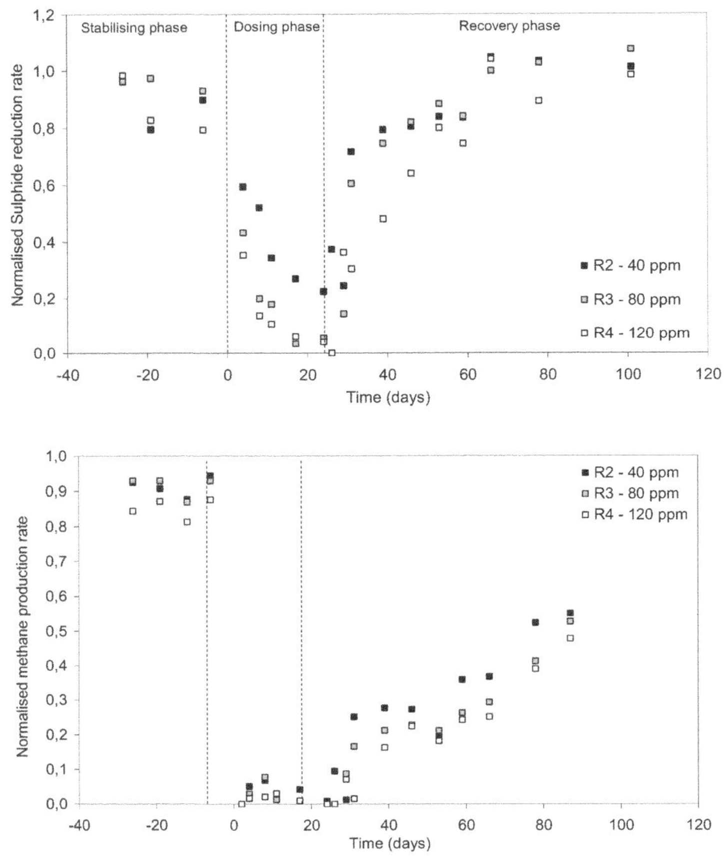

Figure 15.25 Normalized CH_4 production rate in nitrite-dosed reactors: R2 (○), R3 (□), and R4 (Δ), relative to the control reactor R1 (i.e. the activity measured for R1 on each measurement day was considered 100%, to eliminate the rate variation caused by changes in wastewater quality). Nitrite addition was initiated on Day 0 and finished on Day 24.

A field trial was conducted to test intermittent nitrite dosing as a means for sulphide and CH_4 control in a real sewer line, UC09, located in Gold Coast, Australia. As described in a previous section, UC09 has a length of 1080 m and a diameter of 150 mm. It receives primarily domestic wastewater with an average daily flow of approximately 200 m^3. The hydraulic retention time of sewage in the rising main varied between 1.7 and 5.7 hours during the period of the study.

Nitrite solution was dosed into the UC09 wet well manually during daytime only (8:00 a.m. to 7:00 p.m.) over three consecutive days. Before each pumping event, nitrite was added to the wet well, resulting in a concentration in sewage of 100 mg-N/l, a level determined based on the lab-scale reactor test results (Jiang et al., 2010a, see also Figure 15.25). The dosage was stopped after three days. Therefore, the actual dosage time was 33 hours over a 3-day period.

Prior to nitrite dosing, three measurement campaigns were conducted to monitor CH_4 production in the rising main over a period of 7 days. Each campaign involved sampling wastewater hourly at both the pumping station wet well and 828 m downstream of the pumping station, for a period of 3–6 hours. The samples were analyzed for dissolved CH_4 concentrations using the method described in a previous section (5.5.1). Similar measurement campaigns were also conducted 1, 4, 5, 10 and 13 weeks after nitrite dosing.

Figure 15.26 shows the CH_4 concentrations at the pumping station wet well and 828m downstream, where a sampling point was installed. Complete suppression of CH_4 production by nitrite addition was observed, as indicated by the nearly identical CH_4 concentrations in the pumping station and at 828m. One month after terminating nitrite dosage, CH_4 concentration at 828m remained at a level similar to that measured in the wet well, indicating that the sewer biofilm ceased to produce CH_4 in this period. The activity increased only marginally in the following two months.

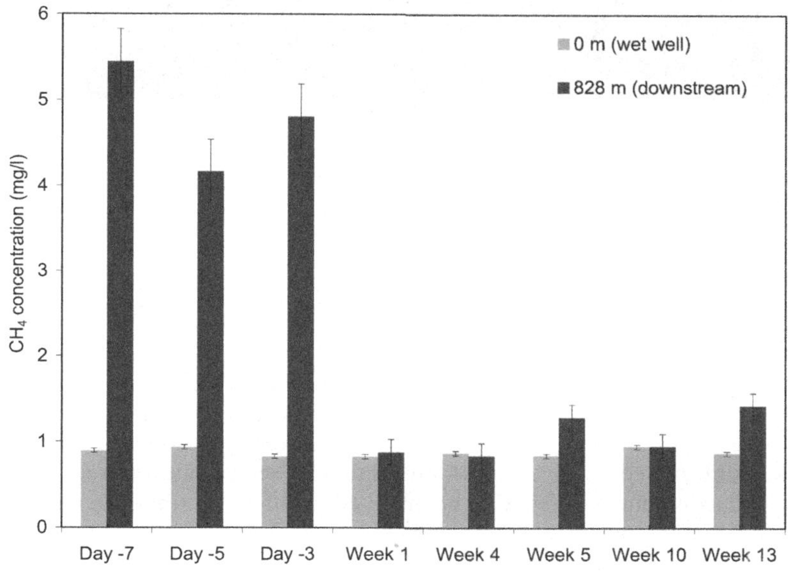

Figure 15.26 Daily average CH_4 concentration at 828m and at the pumping station wet well. Day -7, -5, and -3 were during the baseline period. Nitrite was added intermittently at 100 mgN/l during Day 0–2 for a total dosing time of 33 hours over a 3-day period. Week 1 to 13 imply 1–13 weeks after the dosage. Methane concentrations were based on manual sampling and offline measurements with the method previously described (3–6 samples were taken on each sampling day). The error bars shown are standard errors (n = 3–6).

In general, the field trial confirmed lab study results that nitrite has a long-term toxic effect on methanogens (and sulphate reducing bacteria – data not shown) in anaerobic sewer biofilms. Both the field and laboratory results collectively suggest that nitrite could be applied intermittently to achieve sulphide and CH_4 control in sewers. However, given that nitrite is one of the factors that correlate with N_2O production, the impact of upstream nitrite addition and dosages on the potential on downstream N_2O emissions needs to be determined.

Jiang et al. (2010b) revealed that the suppression of CH_4 (and sulphide) production was due to a strong biocidal effect of free nitrous acid (FNA), the protonated form of nitrite, on microorganisms. The viable cells in sewer biofilms were found to decrease substantially from approximately 80% prior to nitrite dosage to 5–15% after 6–24 hour treatment at FNA levels above 0.2 mg-N/l (equivalent to 80 mgNO_2^--N/l at pH 6.0).

An example of the live and dead staining prior to and after FNA treatment is given in Figure 15.27.

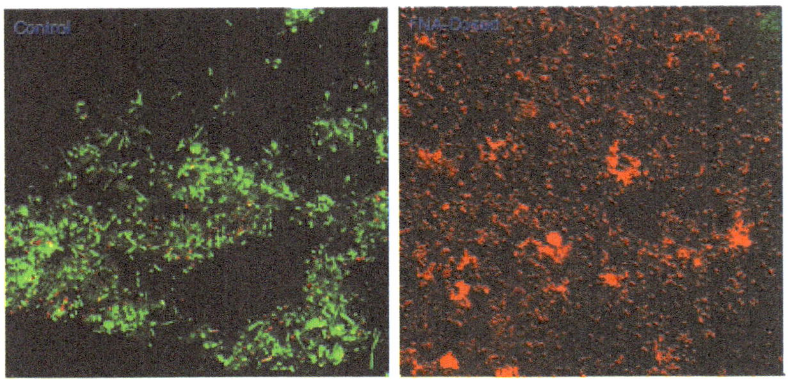

Figure 15.27 Confocal Laser Scanning Microscopy images of homogenized sewer biofilms stained with Baclight® LIVE/DEAD staining kit prior to (left) and after (right) FNA treatment. Green and red cells are intact and damaged cells, respectively.

Iron salt addition

With a similar approach, Zhang et al. (2009) studied the impact of ferric chloride addition on microbial activities of anaerobic sewer biofilms.

Two lab-scale rising main sewer systems fed with real sewage were operated for 8 months. One received Fe^{3+} dosage at 15 mg per l of wastewater (experimental system) and the other was used as a control (no Fe^{3+} dosage). In addition to precipitating sulphide from bulk water, Fe^{3+} dosage was found to significantly inhibit sulphate reduction and CH4 production by sewer biofilms, measured through batch tests. The sulphate reduction rate was found to be inhibited by 40–60%, while the methanogenic activity was inhibited by 50–80%. The rate data were supported by significantly lower CH4 concentrations at the end of the experimental rising main system (for more details, see Zhang et al., 2009). The mechanisms responsible for the inhibition observed are yet to be fully understood.

Oxygen and nitrate addition

The injection of pure oxygen/air and nitrate, which is also commonly used for sulphide control in sewers, may also reduce CH_4 production due to elevated redox conditions. Oxygen may also be toxic to methanogens. However, the detailed effect is yet to be revealed despite that Mohanakrishnan et al. (2009) produced some preliminary evidence showing the absence of CH_4 production in a nitrate-receiving sewer system. Detailed studies on the effect of oxygen and nitrate addition on CH_4 production in sewers are currently underway at The University of Queensland.

15.6 METHANE EMISSION FRANCE

15.6.1 Emission of CH_4

The emission of CH_4 that was measured at the four plants is presented in Table 15.12.

Table 15.12 Emission of CH_4 at four WWTPs in France

WWTP	% CH_4/ of $COD_{Influnet}$
WWTP 1	<0,01%
WWTP 2	<0,01%
WWTP 3	0,04%
WWTP 4	0,03

In the four WWTPs, CH_4 represented 2–10% of the total direct emissions from biological reactors.

15.6.2 Origin CH_4 emission

The emission of CH_4 was very low in activated sludge reactors. Higher emissions can be found at pre-treatment or primary settling tanks but they were not quantified.

15.7 METHANE EMISSION UNITED STATES OF AMERICA

15.7.1 Collection system phase 1: Emission factors

Figure 15.28 shows the winter and summer CH_4 concentration readings at the lift stations from the phase 1 data collection.

Table 15.13 provides CH_4 mass emissions from all three of the LCC series lift stations for comparison. The following observations are of interest:

- The reported emissions for LCC-1 are by far the largest of these three stations that convey approximately the same flow. This high value is likely a function of the continuous ventilation/flux and continuous ventilation methodology than any other factor.
- LCC-3, on the other hand, has the lowest reported emissions while receiving all of its flow from a force main (unlike the other two stations which are gravity fed) which is understood to increase the anticipated CH_4 production. The lower value is likely due to the uncovered fore bay and significant dilution; the other two stations have covered fore bays which limit dilution/dispersion of produced CH_4.

Table 15.13 Comparison of CH_4 emissions from LCC pumping stations.

Lift station name	Estimated CH_4 emissions (kg CH_4/year)	Calculation method	Notes / Comments
LCC-1	4,264	Mechanical ventilation rate and concentration in the air discharge header	Enclosed wetwell with positive ventilation
LCC-2	1,884	"breathing cycle" method	Covered wetwell with no ventilation, some CH_4 could freely escape through openings for the screens
LCC-3	413	"breathing cycle" method	Completely open wetwell

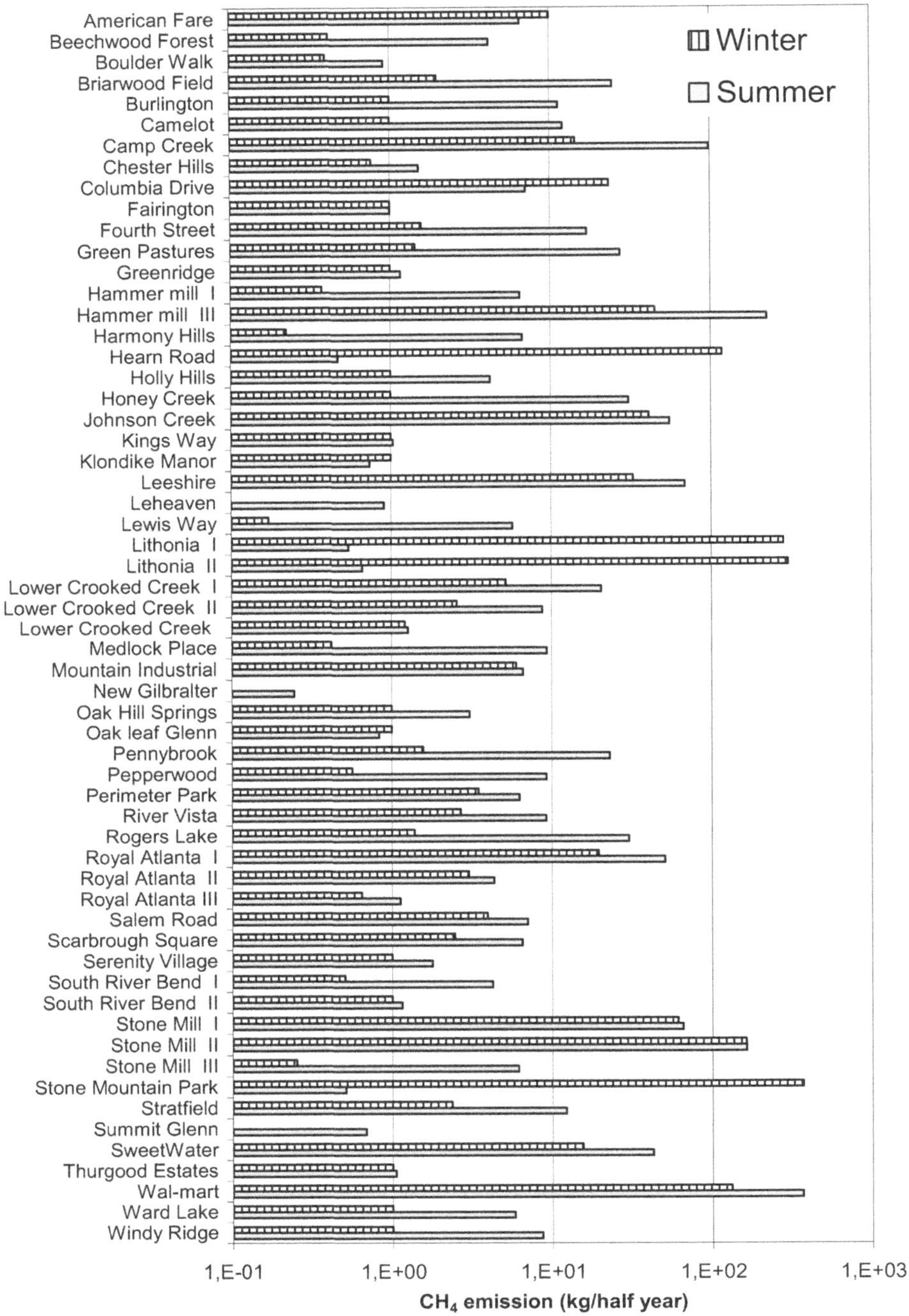

Figure 15.28 Instantaneous CH_4 readings during winter and summer monitoring at 59 lift stations in DeKalb County, Georgia, USA (note x-axis is logarithmic)

15.8 METHANE EMISSION THE NETHERLANDS

15.8.1 Emission of CH_4

The emission of CH_4 as found during the measurements at three Dutch WWTPs is summarised in Table 15.14.

Table 15.14 Emission of CH_4 from three WWTPs.

WWTP	Emission ($kgCH_4/kgCOD_{influent}$)
Papendrecht	0.0087
Kortenoord	0.0053
Kralingseveer October	0.012
Kralingseveer February	0.008
Emission factor NIR*	0.007 without sludge digestion
	0.0085 with sludge digestion

*National Inventory Report.

Based on Table 15.4 the following observations can be made:

- The highest emission of CH_4 occurred at WWTP Kralingseveer in October. This could be related to the presence of a sludge digester at this WWTP where at Papendrecht and Kortenoord a sludge digester is absent.
- In February the emission at Kralingseveer was lower than in October. This could be related to the temperature of the water which was 19°C in October and around 10 °C in February. At lower temperatures less CH_4 will (probably) be produced and more CH_4 will be dissolved. For this reason CH_4 could have left the WWTP via sludge or the effluent. However, the concentration of soluble CH_4 in both streams was not determined.
- The emission factors found for CH_4 are in some case higher in other cases lower than the emission factor currently used to estimate the emission of CH_4 from WWTPs.

15.8.2 Origin of CH_4 emission

The total CH_4 emission was based on measurements at several locations at the three WWTPs. For Papendrecht and Kortenoord the origin of the CH_4 emission is presented in Figure 15.29. Based on Figure 15.29 it can be observed that in Papendrecht the inlet works and aeration tank contributed most to the emission of CH_4. The contribution from the inlet works was 45%, the aeration contributed 32%.

The emission of CH_4 at WWTP Kortenoord primarily occurred from the inletworks, selector and aeration tank, the contribution from the sludge handling was minimal. The contribution from the inlet works varied between 34 and 57%, the contribution from the selector varied between 21–41%, the contribution from the sludge handling remained limited with a maximum of 11%.

For both measurement periods at Kralingseveer the origin of the CH_4 emission is presented in Figure 15.30.

At WWTP Kralingseveer the ventilated air from all anaerobic parts is treated separately in a compost filter. After this treatment the air is reused in the aeration tanks 1 and 2.

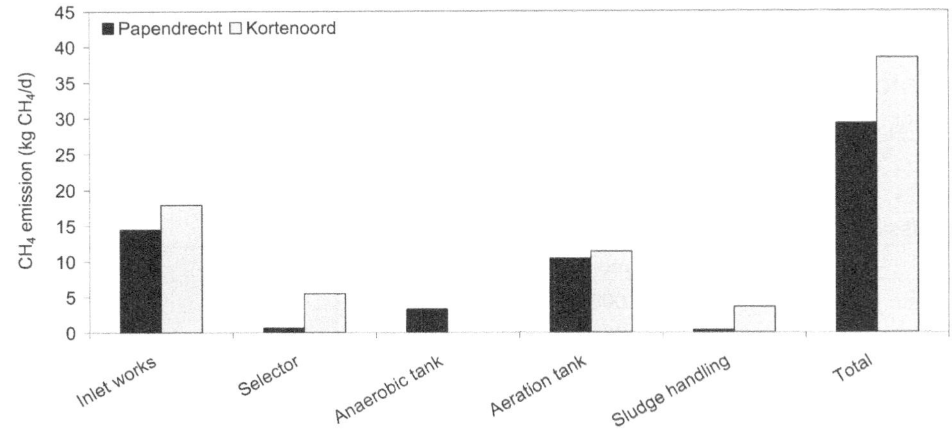

Figure 15.29 Overview of the CH_4 emission at WWTP Papendrecht and Kortenoord.

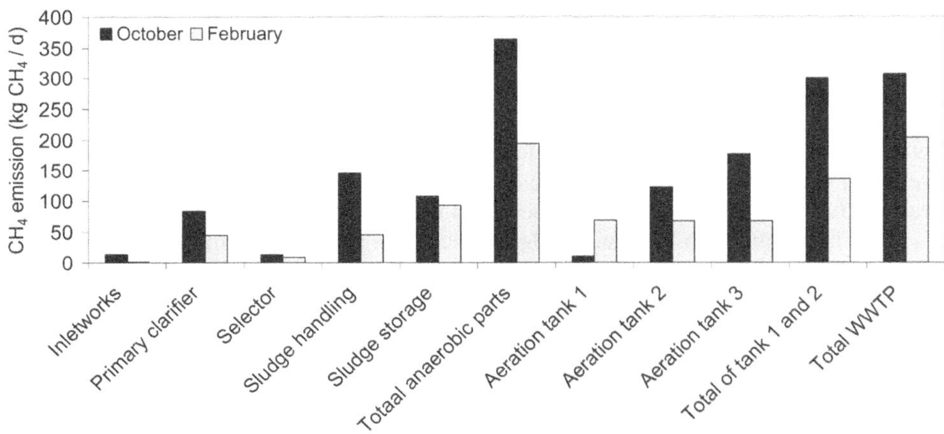

Figure 15.30 Overview of the CH_4 emission at WWTP Kralinseveer during the measurement campaign in October (2008) and February (2009).

For this reason the total CH_4 emission after the anaerobic parts is presented as well in Figure 15.30.

From Figure 15.30 it can be observed that the emission of CH_4 from all anaerobic parts is larger than the total CH_4 emission from aeration tanks 1 and 2. This implies that some CH_4 is lost or converted during treatment in the compost filter or in the old aeration tanks. The loss of CH_4 in October was 92 kg CH_4/d and for February the loss was 59 kg CH_4/d.

The main parts that contributed to the emission of CH_4 were the primary clarifier, the sludge thickeners, handling and storage. The contribution of the primary clarifier was 23% in both periods. For the sludge thickeners and sludge handling the contribution was 45% in October and 23% in February. Finally the sludge storage contributed 28% in October and 48% in February.

15.9 TOTAL CARBON FOOTPRINT

The contribution of each greenhouse gas to the total carbon footprint of the three WWTPs is presented in Figure 15.31.

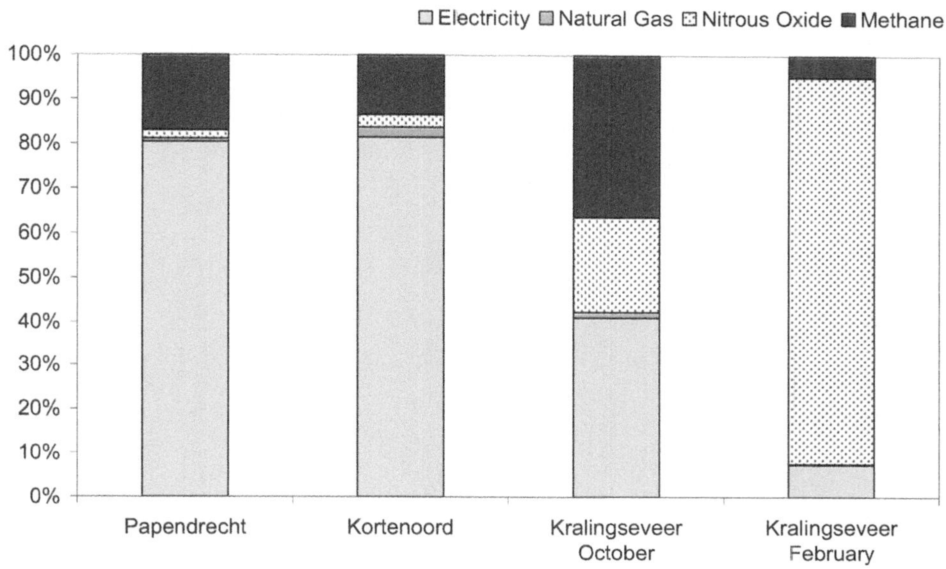

Figure 15.31 Total carbon footprint of WWTP Kortenoord (A; the Netherlands) and WWTP Kralingseveer October (B; the Netherlands)

From Figure 15.31 it can be observed that:

- Electricity forms the major contributor to the total carbon footprint at WWTPs with a low N-sludge load (Papendrecht and Kortenoord).
- At the WWTP (Kralingseveer) with a higher N-sludge load the contribution of N_2O can be substantial.
- The contribution of CH_4 to the total carbon footprint depends on the presence of a sludge digester. In the last case (Kralingseveer) the contribution of CH_4 can be substantial.

Chapter 16
Discussion

16.1 METHODOLOGY

16.1.1 Nitrous oxide emission

The research presented in this report used different methodologies to determine the emission of N_2O from WWTPs. The weak and strong points of these monitoring technologies are summarised in Table 16.1.

Table 16.1 Strong and weak points methodologies to determine N_2O emission from WWTPs.

Methodology	Variability	Origin/process parameters of influence
Liquid phase measurements / mass balance (Australia)	-	+
Gas hood (France/USA[1])	+/-	+
Gas phase covered WWTPs (Netherlands)	+	+/-

[1]Liquid phase measurements were taken as well.

To determine the N_2O emission from uncovered WWTPs different methodologies were developed. In Australia a method was developed based on liquid phase measurements for individual reactor zones. Based on mass transfer coefficients a complete mass balance over the different zones and the complete WWTP could be made. In this way the emission of N_2O was determined, but it was also possible to differentiate between N_2O emission and generation. However, this method does not allow for sufficient insight in the variation of N_2O emission as the method is based on grab samples.

The methodologies developed in the USA and France were based on gas hoods measurements. Those gas hoods were placed at the different zones of a WWTP and in case of the USA, the gas hoods measurements were combined with liquid phase measurements. By placing the gas hoods at the different zones of a WWTP and combine those with the liquid phase measurement it was also in the USA possible to differentiate between N_2O emission and generation. On the other hand, the gas-hood measurements are limited by the dimensions of the hood itself. Additionally, heterogeneities in the emission gas flow rate could also contribute to the variability in the actual emissions, which needs to be considered. To calculate the total emissions load from a plant, both the N_2O concentrations and advective gas flow must linearly scaled up from the hoods to the entire activated sludge bioreactor. Alternately, multiple measurements need to be done at different locations in the activated sludge reactor to capture the spatial variability in emissions.

The methodology used in the Netherlands was based on the analyses of the total off gas of covered WWTPs. The advantage of this method is that it captures the total emissions given the variability of N_2O emission in time and space, but with this methodology it is more difficult to differentiate between N_2O emission and generation.

To finally estimate and control the emission of N_2O from WWTPs, the use of all available methods will be required.

16.1.2 Methane emission

In the research performed in Australia, no CH_4 emission data were calculated. In the USA, the emission of CH4 was calculated from unventilated lift stations. For this gas flow needed to be determined. This was done via the active volume of each lift station in combination with the pumping rate of each of the pumps. It is hereby assumed that the amount of air breathed in and exhausted out during each pumping cycle is approximately equal to the active volume inside the wet well or the volume of raw sewage pumped during each cycle. However, the methodology used to determine the gas flow is subject to a number of limitations, namely:

- Dispersion of CH_4, which is lighter than air, is not accounted for. Many stations cycle only a few times per day and those stations likely lose a considerable portion of the evolved CH_4 to the atmosphere in between cycles. This phenomenon would result in under-reporting of actual emissions.
- Dispersion effects can be very significant at some of the larger pumping stations, like LCC-3. LCC-3 has a large, uncovered forebay with significant interchanges of outside air. On calm days the forbay air changes could be one to five times per hour and could increase to as much as 10 or 30 times per hour on windy days. This limitation, combined with the fact that these stations have the highest volume and mass throughput which should produce more CH_4, could represent a very significant source for under-reporting.
- During a pumping cycle, a significant volume of outside air is drawn into the wet well. If a sampling event occurred soon after such a dilution, the applied concentrations could under-report the actual CH_4 emissions.

These limitations show that there is a need for a good strategy to measure loads of CH4 emitted to the air from sewerage systems.

16.2 NITROUS OXIDE EMISSION

16.2.1 Emission

The emission of N_2O as measured at all the different WWTPs in Australia, France, USA and the Netherlands is summarised in Table 16.2.

Table 16.2 Overview N_2O emission at the surveyed WWTPs around the world (values presented in bold are reported outliers).

Country	WWTP	Sample round	Emission (kg N_2O/kg $TKN_{influent}$)	Emission (kg N_2O-N/kg $TKN_{influent}$)
Australia	1 – Ox. Ditch	1.	0.005	0.003
		2.	0.003	0.002
		3.	0.006	0.004
		4.	0.003	0.002
	2 – Johannesburg	1.	0.016	0.010
		2.	0.027	0.017
		3.	0.011	0.007
	3 – SBR	1.	0.009	0.006
		2.	0.011	0.007
		3.	**0.050**	**0.032**
	4 – MLE (1)	1.	0.019	0.012
	5 – MLE (2)	1.	0.014	0.009
		2.	**0.047**	**0.030**
		3.	0.006	0.004
		4.	**0.055**	**0.035**
	6 – MLE (3)	1.	**0.176**	**0.112**
		2.	0.008	0.005
		3.	0.005	0.003
	7 – A^2O	1.	0.011	0.007
		3.	0.006	0.004
France	WWTP 1		<0.0002	<0.0001
	WWTP 2		<0.0002	<0.0001
	WWTP 3		0.003	0.0017
	WWTP 4		0.002	0.0011
USA	Separate-stage BNR	1. (15°C)	0.0005	0.0003
		2. (23°C)	0.0002	0.0001
	Four-stage Bardenpho	1. (14°C)	0.0025	0.0016
		2. (23°C)	0.0094	0.006
	Step-feed BNR 1	1. (19°C)	0.025	0.016
		2. (25°C)	0.0097	0.0062

(*Continued*)

Table 16.2 Overview N_2O emission at the surveyed WWTPs around the world (values presented in bold are reported outliers)(*Continued*).

Country	WWTP	Sample round	Emission (kg N_2O/kg $TKN_{influent}$)	Emission (kg N_2O-N/kg $TKN_{influent}$)
	Step-feed non-BNR	1. (17°C)	0.0028	0.0018
		2. (26°C)	0.028	0.018
	Separate centrate[1]	1. (30°C)	0.0038	0.0024
		2. (34°C)	0.0085	0.0054
	Plug-flow 1	1. (11°C)	0.0063	0.004
		2. (23°C)	0.0064	0.0041
	Plug-flow 2	1. (11°C)	0.0097	0.0062
		2. (22°C)	0.0014	0.0009
	MLE 1	1. (26°C)	0.0011	0.0007
	MLE 2	1. (26°C)	0.0009	0.0006
	Step-feed BNR 2	1. (29°C)	0.024	0.015
	Oxidation ditch	1. (19°C)	0.0005	0.0003
	Step-feed BNR 3	1. (24°C)	0.0008	0.0005
Netherlands	Papendrecht	1. (19°C)	0.00063	0.00040
	Kortenoord	1. (19°C)	0.00075	0.00048
	Kralingseveer	1. (18°C)	0.0066	0.0042
		2. (10°C)	**0.096**	**0.061**

[1]Not a typical WWTP.

The results of the emission of N_2O from WWTPs (see Table 16.2) exhibit a great variability among different WWTPs and at the same WWTP during different seasons or during the day. This great variability could be understood from the fact that the formation of N_2O is very complex and can be executed by both autotrophic and heterotrophic bacteria.

Those bacteria show a dynamic response to changing influent loads and process conditions resulting in a variable formation of N_2O in time and place. Because of the observed variability in N_2O emission the use of a generic emission factor to estimate the emission from a specific WWTP is inadequate. To determine the emission of N_2O from a specific WWTP measurements are required.

16.2.2 Origin

The results presented in this report showed that the majority of the N_2O emission occurred from aerated zones. This emission was observed at the transition point between anoxic and aerobic zones (see Figure 15.4 and Figure 15.10) and in complete aerobic zones (see Figure 15.4 and Figure 15.10). The emission of N_2O at the transition point between anoxic and aerobic zones can be the result of:

- Stripping of N_2O formed during denitrification
- N_2O formation by heterotrophic denitrification
- N_2O formation by autotrophic nitrification

Stripping of N_2O formed during denitrification can primarily occur at the transition of anoxic to aerobic zones an example of this can be seen from Figure 15.10. At this transition point between anoxic and aerobic conditions formation of N_2O is possible by both heterotrophic denitrification and autotrophic nitrification. Heterotrophic N_2O emissions can be related to oxygen inhibition as was shown by several authors (Knowles, 1982; Korner and Zumft, 1989; Zumft, 1992; Hanaki et al., 1992). Autotrophic N_2O emissions can be related to the fact that in the presence of non limiting DO and NH_4^+-N concentrations (as occurs at the transition point between anoxic and aerobic zones) N_2O can be formed by autotrophic nitrification as was recently proven by Yu et al. (2010). This process is most likely also responsible for the formation of N_2O under complete aerobic conditions, although the possibility of heterotrophic denitrification within sludge flocs cannot be discounted (Grady et al., 1999).

Although both heterotrophic denitrification and autotrophic nitrification can be responsible for the formation of N_2O, autotrophic nitrification seems to be the major contributor to the total N_2O emission from a WWTP based on the results reported by the USA (Figure 15.10) and France (Figure 15.4). This is supported by the found positive correlations between NO_2^--N, NH_4^+-N and DO concentrations and N_2O emission (see Table 15.4).

16.2.3 Process parameters influence

Process parameters that influenced the emission and formation of N_2O from aerobic zones were found to be NO_2^--N, NH_4^+-N and DO (isolated effect) and NH_4^+-N – NO_2^--N (interactive effect) as could be observed from Table 15.4. All these parameters showed a positive correlation with the emission of N_2O, which means that the emission of N_2O increased as the concentration of the mentioned components increased.

Nitrite is one of the most important factors contributing to N_2O emission from nitrification as was proven by Beaumont et al. in several papers (2004a, 2004b and 2005).

From this the correlation of NO_2^--N with aerobic zone N_2O emissions is consistent with basic biochemical principles. However, it should be noted that the inhibitory effect of nitrite does not further increase at levels above ~2 mg NO_2^--N/l.

The positive correlation between NH_4^+-N and DO with N_2O emission are in congruence with work of Yu et al. (2010) on N_2O formation by autotrophic nitrification. In this work it was proven that abrupt increases in DO concentrations in the presence of NH_4^+-N can lead to transient accumulation of NO_2^--N, which in turn result in autotrophic NO and N_2O generation (Yu et al., 2010). Furthermore, it was shown that the magnitude of autotrophic nitrification driven N_2O emission also directly depends on the NH_4^+-N concentrations. The positive interactive correlation between NH_4^+-N and NO_2^--N concentrations and N_2O emissions suggests high N_2O emissions from aerobic zone locations, with simultaneously high concentrations of both NH_4^+-N and NO_2^--N. This interactive correlation also points to autotrophic N_2O generation mechanisms, since both NH_4^+-N and NO_2^--N are co-substrates in autotrophic denitrification by nitrifying bacteria (Beaumont et al., 2005, 2004a, 2004b).

Formation and emission of N_2O from anoxic zones was only found to be influenced by the NO_2^--N – DO concentration (interactive effect). This positive correlation of NO_2^--N – DO concentrations with N_2O emissions is consistent with known mechanisms of denitrification-related N_2O production. These mechanisms were proven by several authors. DO inhibition during denitrification was proven by Tallec et al. (2008) and Park et al. (2000); NO_2^-- N inhibition was proven by Korner and Zumft (1989), von Schultesss et al. (1995) and Zumft (1997). Inhibition of denitrification by high DO concentrations could also lead to NO_2^--N build up, indirectly leading to N_2O emissions (Hanaki et al., 1992; Zhou et al., 2008).

16.2.4 Implications of gained knowledge

Based on the above, it becomes clear that at increasing levels of NH_4^+-N, NO_2^--N and DO, the formation of N_2O increases in aerobic zones. This implies that if high concentrations of these constituents can be avoided in practice, the risk of N_2O emission will be reduced. These high concentrations can be avoided in:

- systems that approach "ideal" well-mixed conditions (i.e. high recycle rates). In this way concentrations of intermediates from nitrification-denitrification including NO_2^--N and NO are diluted, thereby reducing their inhibitory effect (Casey et al., 1999a,b). This is confirmed by the general results from all studies that WWTPs with the highest recycle rates and with the lowest TN effluent concentrations belonged to the WWTPs with the lowest N_2O emission.
- systems that avoid over – aeration i.e. have a rapidly responding DO control systems. In this way high levels of DO in aerobic and anoxic zones can be avoided
- systems that are equipped with flow equalization. Such systems minimize the peaking factor of influent nitrogen (ammonium) loading to the activated sludge system.

With the presented insight, it has been possible to make a first estimate on the risk level of N_2O emissions from a specific WWTP. This estimation can be based on the risk matrix presented in Table 16.3.

Table 16.3 Risk matrix to determine risk level of N_2O emission.

Risk on N_2O	High	Medium	Low
Parameter			
Effluent total organic nitrogen (mg/l)	>10	5–10	<5
Range in N-concentration in plant	H	M	L
Load variations	H	M	L
Maximum NO_2^--N conc. anywhere in plant	>0.5*	0.2–0.5	0.2

* Risk does not increase at NO_2^--N concentrations higher than 2 mg N_2O-N/l.

In summary, it can be concluded that in systems with extensive nitrogen removal the formation of N_2O is greatly minimized. In other words there is no conflict between water quality and air quality; rather they go hand in hand. This implies that systems that are not designed for (complete) nitrogen removal exhibit a high risk of N_2O emission. At temperatures of 20°C, an aerobic SRT of 2–3 days is already sufficient to achieve nitrification. In these high-loaded systems, nitrite easily accumulates resulting in greater N_2O formation.

16.2.5 Future research

The results of the presented research performed in Australia, France, the USA and the Netherlands already clearly identified key process parameters that influence the emission of N_2O from WWTPs. This knowledge presents a good starting point for defining mitigation strategies to reduce the emission of N_2O. Future research should first give better insight in the variability of N_2O emissions over prolonged times in order to develop proper sampling protocols for emissions of N_2O from treatment plants. A better understanding of the processes leading to nitrite formation in treatment plants can lead to mitigation strategies for N_2O emissions. Besides activated sludge systems, biofilm processes should be monitored for potential emissions. A better insight in N_2O production in natural systems can improve decisions on implementing full nitrogen removal as a mitigation strategy for N_2O emissions from treatment systems as well as natural systems.

16.3 METHANE EMISSION

16.3.1 Sewers

The measurements performed in Australia and the USA show that substantial amounts of CH_4 can be formed and emitted from sewer systems. In addition to this substantial amounts of CH_4 were measured from the inlet works at the WWTPs in the Netherlands. With this knowledge, the contribution of the sewer system can not be neglected as it is dismissed in the current IPCC guidelines (IPCC 2006a). However, to be able to make a good estimate of the contribution of sewers to the total CH_4 emission from wastewater collection and treatment systems more data are required and a good strategy to measure CH_4 loads is necessary (refer to section 16.1.2). The data collected to date are far from adequate for the development of reliable accounting guidelines, and for the development, calibration and validation of CH_4 emission models. It is of utmost importance that such data are collected from sewer networks (rising mains and gravity sewer) around the world under different climate conditions, and from networks collecting and conveying various types of wastewaters. Both liquid and gas phase data are urgently needed.

The data obtained in Australia suggest that the wastewater composition, HRT and the dimensions (Area / Volume ratio) are parameters that can influence the level of CH_4 formation from a rising main.

The results from the lift stations in the USA (Figure 15.28) showed that the concentration of CH_4 in the gas phase was in almost all cases (80%) higher in summer than in winter. This indicates that the temperature is an important parameter determining the formation of CH_4 in sewers. This might also explain relative high CH_4 levels in Australia (although this could also result from easy degradable nature of wastewater).

Finally the gas phase measurements (%LEL) in Australia indicate that the composition of wastewater is another parameter that determines the formation of CH_4 in sewers. The results of these field measurements were confirmed by results from laboratory measurements performed by Sudarjanto et al. (2010b). In these laboratory experiments the addition of a certain volume by percent of brewery wastewater to raw sewage lead to an increase in the CH_4 production rate compared to a control reactor. Interestingly in another experiment described by Sudarjanto et al. (2010b) it was found that the addition of a certain volume by percent of dairy waste to raw sewage did not lead to any increase in CH_4 production. These results suggest that each trade waste should be assessed separately, but that the composition of the wastewater is a parameter to taken into account for the formation of CH_4 in sewers.

16.3.2 Mitigation strategies

Several strategies used for sulphide control in sewers have been found to reduce CH_4 formation (see section 15.5.5). However these strategies may not be the most cost-effective methods for CH_4 control. Given the high sensitivity of methanogens to environmental changes, more cost-effective strategies may be developed. The addition of nitrite (free nitrous acid) is a very promising technology in this respect, but might easily lead to N_2O formation.

16.3.3 Wastewater treatment plants

Based on the results from the Netherlands it was found that the current emission factors used in the Netherlands are for the present useful to estimate the emission from WWTPs. However, the amount of measurements was limited and in general measurements at several WWTPs should be performed to confirm existing emission factors used per country.

The results from the measurements at the WWTPs without sludge digestion (Figure 15.29) showed that CH_4 emission primarily occurred from the inlet works and the aeration tanks. The high emission from the inlet works strongly indicates that the CH_4 emitted originates from the sewer as due to the very short HRT formation of CH_4 in the inlet works can be excluded. Furthermore, the wastewater to both WWTPs is transported via rising mains in which under anaerobic conditions CH_4 formation is possible as was shown by the results from Australia. The rather large contribution of the aeration tank to the total CH_4 emission is rather strange as under anoxic and aerobic conditions CH_4 formation is very unlikely. Part of the CH_4 emitted from the aeration tank could have been formed in the anaerobic tank or anoxic tank upstream the aeration tanks. However, this could not be confirmed due to the lack of measurements from the liquid phase. Furthermore, it is known that limited amounts of methanogens are present in activated sludge (Gray, et al., 2002 and Lens et al., 1995). Based on these considerations it is thought that the CH4 emitted from the aeration tank mainly originates from the sewer system.

At the WWTP with sludge digestion the highest relative CH_4 emission was found from the sludge handling and storage sites (in total ~50%). This could be understood from the fact that these sites handled sludge from the digester. The contribution from the sewer was in this case smaller (~25%).

From the results at the WWTP it is shown that process units that are known for their sulphur related odours (inlet works, sludge digesters) also emit CH_4.

16.4 TOTAL CARBON FOOTPRINT WWTP

As a first indication on the possible contribution of N_2O and CH_4 emission to the total carbon footprint of a WWTP, the result in the Netherlands could be used as an example.

In the case studies in the Netherlands, the specific emissions of N_2O and CH_4 were determined at the same time. Together with the data on the related consumption of electricity and natural gas, it was possible to calculate a carbon footprint of three WWTPs. To determine the carbon footprint, all sources were converted to CO_2 equivalents[1]. The results in the Netherlands indicated that the emission of CH_4 and N_2O can significantly contribute to the total carbon footprint of a WWTP. This contribution can vary from 2% to almost 90% of the carbon footprint under extreme conditions for N_2O and 5–40% for CH_4. One should be aware that these numbers are specific for the Netherlands. In any other country, these numbers can differ greatly as there exist a great variation in the way wastewater and sludge is handled as well as the specific composition of the energy mix used.

Furthermore these numbers can significantly differ depending on how the boundaries are set around the analysis. In case of the analysis performed for the three Dutch WWTPs the contribution of e.g. chemical use, and sludge incineration were not accounted for.

[1] It should be noted that the conversion numbers are country specific and do depend on the used energy mix (i.e. brown coal versus wind or solar energy), which is of influence on the total carbon footprint of a WWTP.

Chapter 17

Conclusions and future research

17.1 CONCLUSIONS

17.1.1 Nitrous oxide emission

Based on the research presented in this report the following conclusions can be drawn:

- The emission of N_2O is highly variable. For this reason the use of generic emission factor to estimate the emission from a specific WWTP is inadequate.
- To determine the emission of a specific WWTP, measurements are required. These measurements should be performed with online monitors and over the operational range of the WWTP.
- Emission of N_2O originates mainly from nitrification.
- Accumulation of nitrite leads to the formation of N_2O in mainly aerobic zones.
- High NH_4^+-N concentrations can lead to the emission of N_2O if nitrification occurs.
- The risk level of N_2O emission can be determined based on the following risk matrix:

Table 17.1 Risk matrix to determine risk level of N_2O emission.

Risk on N_2O	High	Medium	Low
Parameter			
Effluent total organic nitrogen (mg/l)	>10	5–10	<5
Range in N-concentration in plant	H	M	L
Load variations (daily)	H	M	L
Maximum NO_2^--N concentration (mg N/l) anywhere in plant	>0.5*	0.2–0.5	0.2

*Risk does not increase at higher NO_2^--N concentrations

- In addition to the risk matrix, a rapidly responding DO control also contributes in minimising the risk of N_2O emission.

- A good effluent quality (TN < 5 mgN/l) goes hand in hand with a low risk of N_2O emission.
- Non BNR systems with unintentional nitrification will have a high risk of N_2O emission.

17.1.2 Methane emission

Based on the research presented in this report the following conclusions can be drawn:

- Formation and emission of CH_4 from sewers can be substantial.
- Sulphur related odours are good indicators for CH_4 formation.
- Odour mitigation strategies in sewers likely also support a reduced CH_4 formation.
- Methane emission from WWTPs mainly originates from sewers and sludge handling.
- Emissions of CH_4 and N_2O significantly contribute to the total greenhouse gas footprint of WWTPs.

17.1.3 Total carbon footprint

- Emissions of CH_4 and N_2O may significantly contribute to the total greenhouse gas footprint of WWTPs as was based on the results from the Netherlands.

17.2 FUTURE RESEARCH

Based on the outcomes of the research valuable knowledge is gained to estimate and reduce the emission of N_2O and CH_4 from wastewater collection and treatment systems. Future applied research should focus on:

Nitrous oxide

- Variability of N_2O emission such that proper sampling and monitoring programmes can be developed.
- Mitigation strategies (including insight in the relative contribution of autotrophic and heterotrophic processes to N_2O generation) to reduce emission via process design and control;
- Evaluate emission from biofilm based processes;
- Evaluate emission from various receiving aquatic environments.

Methane

- Development of a strategy to measure loads emitted from sewers;
- Build a database for emission measurements from sewers;
- Cost effective mitigation strategies[1];
- Evaluate emission from sludge treatment lagoons.

---

[1]The addition of nitrite seemed to be a strategy that could reduce the formation of methane, however, given that nitrite is one of the factors that correlate with N_2O production, the impact of upstream nitrite addition and dosages on the potential on downstream N_2O emissions needs to be determined as well.

Chapter 18

Acknowledgements

18.1 AUSTRALIA NITROUS OXIDE RESEARCH

This work was funded by the Water Services Association of Australia and The University of Queensland. In-kind support was also provided by United Water International (Adelaide), and the participating water authorities.

The members of the project team for this study were:

Principal authors:	Jeff Foley (PhD student, UQ)
	Professor Paul Lant (UQ)
Fieldwork assistants:	Bill Fuller (Engineer, United Water International)
	Peter Muller (Undergraduate engineer, GHD)
	Con Foley (volunteer)
	Lyndon Clark and Debbie Chua (Water Corporation)
	Ian Johnson (Gold Coast Water)
Analytical and modelling support:	Dr Ursula Werner (UQ)
	Shihu Hu (PhD student, UQ)
	Dr David de Haas (UQ, GHD)
	Dr Keshab Sharma (UQ)
	Dr Albert Guisasola (Universitat Autònoma de Barcelona)
	Dr Beatrice Keller-Lehmann and Jianguang Li (UQ)
	Peng Su and Associate Professor Wei Zhang (Flinders University)
Technical review and stewardship:	Professor Zhiguo Yuan (UQ)
	Professor Jurg Keller (UQ)
	Dr David de Haas (UQ, GHD)
	Dr Paul Rasmussen (United Water International)
	Hamish Reid (South East Water)
	Michael Greg and Daniel Starrenburg (Brisbane City Council)
	Greg Allen (Sydney Water)
	Keith Cadee (Water Corporation)
	Adam Lovell (Science and Sustainability Manager, WSAA)

The authors also thank the plant owners and operators for their cooperation:

Brisbane City Council	Keith Barr, Vijay Jayaraman, Kenny Liew and Grahame Simpson
Fosters Brewery	Charlie Foxall
Gold Coast Water	Shaun Corrie, Ian Johnson and Kelly O'Halloran
Melbourne Water	Kim Daire, Peter Gall, Trevor Gulovsen, Suelin Haynes and Erik Ligtermoet
SA Water/United Water	Tim Kelly, Bridget Kingham, Ian MacKenzie and Dr Paul Rasmussen
Sydney Water	Heri Bustamante, Steven Lin, Mark Ognibene, Dammika Vitanage, Wanxin Wang and Wasantha Wicks
Toowoomba Regional Council	Alan Kleinschmidt and John Paulger
Western Australian Water Corporation	Claire Camplin, Lyndon Clark, Margaret Domurad, Kay Hyde, Pardeep Kumar, Gabrielle O'Dwyer and Leah Rheinberger

18.2 AUSTRALIA METHANE RESEARCH

The following people from the Advanced Water Management Centre at The University of Queensland are thankfully acknowledged for conducting the original research reported in this document on the measurement, understanding, modeling and mitigation of methane formation in sewers: David de Haas, Jeff Foley, Albert Guisasola, Oriol Gutierrez, Guangming Jiang, Jurg Keller, Paul Lant, Janani Mohanakrishnan, Keshab Sharma, Gatut Sudarjanto and Lishan Zhang.

The Australian Research Council (ARC), the Water Services Association of Australia (WSAA) and numerous members of the Australian water industry provided financial support to the research presented in this report through ARC projects LP0454182, ARC LP0882016 (http://www.score.org.au/) and the WSAA project Direct Methane and Nitrous Oxide emissions from full-scale wastewater treatment systems.

Dr Trisha Ibrahim from South East Water Limited (SEWL) kindly provided gas-phase methane data measured at one of the sewer lines operated by SEWL. The discussion with Trisha about these results has been very highly valuable.

Mr Shaun Corrie and Mr Ian Johnston from Allconnex Water (formerly Gold Coast Water) assisted with the collection of dissolved methane data from UC09 and CO16. They also provided on-line gas phase methane data measured at CO16.

18.3 UNITED STATES OF AMERICA NITROUS OXIDE RESEARCH

This work was funded by the Water Environment Research Foundation under Project U4R07 and the Climate Change program. The members of the project team for this study were:

Principal authors:	Kartik Chandran, Ph.D. (Columbia University)
Field researchers:	Joon Ho Ahn, Ph.D. (Columbia University)
	Sungpyo Kim, Ph.D. (Columbia University)
	Hongkeun Park, Ph.D. (Columbia University)
	Brian Rahm, Ph.D (Columbia University)
Laboratorium researchers	Ran Yu, Ph.D. (Columbia University)
	Huijie Lu, Ph.D. (Columbia University)

Technical Review and Direction:	Greg Adams, P.E. (Los Angeles County Sanitary District) Eugenio Geraldo, Ph.D. (American Water) William Hiatt, Ph.D., P.E. Sudhir Murthy, Ph.D., P.E. (DC Water) Denny Parker, Ph.D. P.E. (Brown and Caldwell) Peter Strom, Ph.D. (Rutgers University) Jeff Weiss, Ph.D. (Los Angeles County Sanitary District)

18.4 UNITED STATES OF AMERICA METHANE RESEARCH

This work was funded by the Water Environment Research Foundation under Project U2R08 and the Climate Change program. The members of the project team for this study were:

Principal authors:	John Willis, P.E. (Brown and Caldwell)
Field assistants:	Anup Shah (Brown and Caldwell)
Technical Review and Direction:	Bob Forbes, PE (CH2M-Hill) Eugenio Geraldo, Ph.D., (American Water) Catherine O'Connor, (Metropolitan Water Resources District of Greater Chicago) Eliza Jane Whitman, PE (Parsons) Patrick Wootton, (GE)

18.5 THE NETHERLANDS

The authors acknowledge the contribution of Kyra Schipper, Pieter Vijfhuizen in the collection of field data. The Steering Committee of the project, Alex Sengers, Klaas Appeldoorn, Frank Brandse and Hardy Temmink are thanked for their constructive input during the project. The plant operators at Kralingseveer, Kortenoord and Papendrecht WWTP are also acknowleged for their hospitality and co-operation.

Chapter 19

References

Ahn, J.-H., Kim, S., Pagilla, K., Katehis, D. and Chandran, K., 2009, Spatial and temporal variability in N_2O generation and emission from wastewater treatment plants. Nutrient Removal Conference. Washington D.C., Water Environment Federation

Alberto M. C. R., Arah J. R. M., Neue H. U., Wassman R., Lantin R. S., Aduna J. B. and Bronson K. F. 2000, A sampling technique for the determination of dissolved methane on soil solution. Chemosphere: Global Change Science (2), 57–63.

Anderson, I. C., Levine, J. S., 1986, Relative rates of nitric oxide and nitrous oxide production by nitrfiers, denitrifiers and nitrate respirers. Applied Environmental. Microbiology., 51, 938–945.

Anderson, I. C., Poth, M., Homstead, J., Burdige, D., 1993, A comparison of NO and N_2O production by the autotrophic nitrifier Nitrosomonas europaea and the heterotrophic nitrifier Alcaligenes faecalis. Applied Environmental. Microbiology., 59, 3525–3533.

Andrews, J. Chambers, B., Davey, A., Galletti, S., Hobson, J., Hunt, D., Thorman, R. and Walker, I., 2009, Carbon Accounting in the Water Industry; non-CO_2 Emissions, UKWIR, London

Beaumont, H. J. E., Lens, S. I., Reijnders, W. N. M., Westerhoff, H. V. and Van Spanning, R. J. M. 2004a, Expression of nitrite reductase in Nitrosomonas europaea involves NsrR, a novel nitrite-sensitive transcription repressor. Molecular Microbiology, 54, 148–158.

Beaumont, H. J. E., Van Schooten, B., Lens, S. I., Westerhoff, H. V. and Van Spanning, R. J. M. 2004b, Nitrosomonas europaea Expresses a Nitric Oxide Reductase during Nitrification. Journal Bacteriology, 186, 4417–4421.

Beaumont, H. J. E., Lens, S. I., Westerhoff, H. V. and Van Spanning, R. J. M. 2005, Novel nirK cluster genes in Nitrosomonas europaea are required for NirK-dependent tolerance to nitrite. Journal Bacteriology, 187, 6849–6851.

Benthum van, W. A. J., Garrido, J. M., Mathijssen, J. P. M., Sunde, J., van Loosdrecht, M. C. M. and Heijnen, J. J., 1998, Nitrogen removal in intermittently aerated biofilm airlift reactor. Journal of Environmental Engineering (Reston, Virginia) 124 (3), 239–248.

Burgess, J. E., Colliver, B. B., Stuetz, R. M. and Stephenson, T., 2002, Dinitrogen oxide production by a mixed culture of nitrifying bacteria during ammonia shock loading and aeration failure. Journal of Industrial Microbiology and Biotechnology 29 (6), 309–313.

Capela, S., Roustan, M. and Heduit, A., 2001, Transfer number in fine bubble diffused aeration systems. Water Science and Technology 43 (11), 145–152.

Casey, T. G., Wentzel, M. C. and Ekama, G. A., 1999a, Filamentous organism bulking in nutrient removal activated sludge systems – Paper 10: Metabolic behaviour of heterotrophic facultative aerobic organisms under aerated/unaerated conditions. Water Sanitation 25 (4), 425–442.

Casey, T. G., Wentzel, M. C. and Ekama, G. A., 1999b, Filamentous organism bulking in nutrient removal activated sludge systems Paper 9: Review of biochemistry of heterotrophic respiratory metabolism. Water Sanitation 25 (4), 409–424.

Chung, Y.-C. and Chung, M.-S., 2000, BNP test to evaluate the influence of C/N ratio on N_2O production in biological denitrification. Water Science and Technology 42 (3–4), 23–27.

Cleemput van, O. and Baert, L., 1998, Subsoils: chemo- and biological denitrification, N_2O and N_2 emissions., Nutrient Cycling in Agroecosystems, 52 (2–3), 187–194.

Colliver, B. B. and Stephenson, T., 2000, Production of nitrogen oxide en dinitrogen oxide by autotrophic nitrifiers, Biotechnology Advances 18, pp. 219–232.

Czepiel, P. M., Crill, P. M. and Harriss, R. C., 1993, Methane emissions from municipal wastewater treatment processes, Environmental Science Technology, 27, 2472–2477

Czepiel, P. M., Crill, P. M. and Harriss, R. C., 1995, Nitrous Oxide emissions from municipal wastewatertreatment, Environmental Science Technology, 29, 2352–2356.

DCC, 2008b, National Greenhouse and Energy Reporting (Measurement) Technical Guidelines, 1.1 edition. Department of Climate Change, Commonwealth of Australia, Canberra.

Dean, J. A. (ed) 1992, Lange's Handbook of Chemistry, McGraw-Hill, Inc., New York.

Dudley, J., 1995, Mass transfer in bubble columns: A comparison of correlations. Water Research 29 (4), 1129–1138.

Envirosim, 2007, BioWin Process Simulator, v.3.0, Envirosim Associates Ltd, Flamborough, Ontario

Ferrell, R. T. and Himmelblau, D. M., 2002, Diffusion coefficients of nitrogen and oxygen in water. Journal of Chemical and Engineering Data 12 (1), 111–115.

Foley, J. and Lant, P., 2008, Fugitive Greenhouse Gas Emissions from Wastewater Systems: WSAA Literature Review No.01. Water Services Association of Australia, Melbourne.

Foley, J. and Lant, P., 2009, Direct Methane and Nitrous oxide emissions from full-scale wastewater treatment systems, Occasional paper No.24, Water Service Association of Australia.

Foley, J., de Haas, D., Hartley, K., Yuan, Z., Lovell, A. and Lant, P. (2009) Field measurements and simple model development for estimating greenhouse gas emissions. *Wat. Sci. Tech.* 60 (11), 2863–2971.

Garcia-Ochoa, F. and Gomez, E., 2009, Bioreactor scale-up and oxygen transfer rate in microbial processes: An overview. Biotechnology Advances 27 (2), 153–176.

Garrido, J. M., Moreno, J., Mendez-Pampin, R. and Lema, J. M.,1998, Nitrous oxide production under toxic conditions in a denitrifying anoxic filter. Water Research 32 (8), 2550–2552.

Gejlsbjerg, B., Frette, L. and Westermann, P., 1998, Dynamics of N_2O production from activated sludge, Water Research, volume 32, issue no 7, pp. 2113–2121.

Gillot, S., Capela-Marsal, S., Roustan, M. and Heduit, A., 2005, Predicting oxygen transfer of fine bubble diffused aeration systems-model issued from dimensional analysis. Water Research 39 (7), 1379–1387.

Grady, C. P. L. J., Daigger, G. T. and Lim, H. C., 1999, Biological Wastewater Treatment., New York, Marcel Dekker.

Gray, N. D., Miskin, I. P., Kornilova, O., Curtis, T. P. and Head, I. M., 2002, Occurrence and activity of *Archea* in aerated activated sludge wastewater treatment plants, Environmental Microbiology 4 (3), pp.158–168.

Guisasola, A., Haas, de D., Keller, J. and Yuan, Z., 2008, Methane formation in sewer systems, Water Research (42) p 1421–1430

Gutierrez, O., Park, D., Sharma, K. R. and Yuan, Z. 2009, Effects of long-term pH elevation on the sulfate-reducing and methanogenic activities of anaerobic sewer biofilms. Water Research 43 (9), 2549–2557.

GWRC, 2011, N_2O and CH_4 Emission from Wastewater Collection and Treatment Systems, State of the Science report, London, United Kingdom, ISBN 978–90–77622–23–0.

Hanaki, K., Hong, Z. and Matsuo, T. 1992, Production of nitrous oxide gas during denitrification of wastewater. Water Science and Technology 26 (5–6), 1027–1036.

Hartley, K. and Lant, P., 2006, Eliminating non-renewable CO_2 emissions from sewage treatrment: An anaerobic migrating bed reactor pilot plant study, Biotechnology and Bioengineering, 95 (3), 384–398.

Hasanen, A., Orivuori, P. and Aittamaa, J., 2006, Measurements of local bubble size distributions from various flexible membrane diffusers. Chemical Engineering and Processing 45 (4), 291–302.

Hvitved-Jacobsen, T. 2002, Sewer processes: microbial and chemical process engineering of sewer networks. CRC Press, USA.

Ibrahim, T., 2010, High LELs in a sewer line receiving industrial wastewaters. Internal Resort. South East Water Limited.

IPCC, 1997. Module 6 – Waste. in Revised 1996 IPCC Guidelines for National Greenhouse Gas Inventories, vol.2 - Workbook. Intergovernmental Panel on Climate Change.

IPCC, 2006a. Ch.6 – Wastewater Treatment and Discharge. in H. S. Eggleston, L. Buendia, K. Miwa, T. Ngara, and K. Tanabe, editors. 2006, IPCC Guidelines for National Greenhouse Gas Inventories, Prepared by the National Greenhouse Gas Inventories Programme, vol.5 – Waste. IGES, Japan.

IPCC, 2006b. Vol.1 – General Guidance and Reporting. in H. S. Eggleston, L. Buendia, K. Miwa, T. Ngara, and K. Tanabe, editors. 2006, IPCC Guidelines for National Greenhouse Gas Inventories, Prepared by the National Greenhouse Gas Inventories Programme, IGES, Japan.

Itokawa, H., Hanaki, K. and Matsuo, T., 2001, Nitrous oxide production in high-loading biological nitrogen removal process under low COD/N ratio condition. Water Research 35 (3), 657–664.

Jiang, G., Gutierrez, O., Sharma, K. R. and Yuan, Z. 2010a, Effects of nitrite concentration and exposure time on sulfide and methane production in sewer systems. Water Research, 44 (14), 4241–4251.

Jiang, G., Gutierrez, O. and Yuan, Z., 2010b Biocidal effect of free nitrous acid on anaerobic wastewater biofilms. Water Research. DOI: 10.1016/j.watres.2011.04.026

Kampschreur, M. J., Tan, N. C. G., Kleerebezem, R., Picioreanu, C., Jetten, M. S. M. and van Loosdrecht, M. C. M., 2008a, Effect of dynamic process conditions on nitrogen oxides emission from a nitrifying culture. Environmental Science and Technology 42 (2), 429–435.

Kampschreur, M. J., Van Der Star, W. R. L., Wielders, H. A., Mulder, J. W., Jetten, M. S. M. and Van Loosdrecht, M. C. M., 2008b Dynamics of nitric oxide and nitrous oxide emission during full-scale reject water treatment. Water Research, 42, 812–826.

Kampschreur, M. J., Temmink, H., Kleerebezem, R., Jetten, M. S. and van Loosdrecht, M. C., 2009, Nitrous oxide emission during wastewater treatment, Water Research 43 (17), pp. 4093–4103

Kester, R. A., De Boer, W. and Laanbroek, H. J., 1997, Production of NO and N_2O by pure cultures of nitrifying and denitrifying bacteria during changes in aeration. Appl. Environmental Microbiology., 63, 3872–3877.

Khudenko, B. M. and Shpirt, E., 1986, Hydrodynamic parameters of diffused air systems. Water Research 20 (7), 905–915.

Kimochi, Y., Inamori, Y., Mizuochi, M., Xu, K.-Q. and Matsumura, M., 1998, Nitrogen removal and N_2O emission in a full-scale domestic wastewater treatment plant with intermittent aeration. Journal of Fermental Bioengineering., 86, 202–206.

Knowles, R., 1982, Denitrification, Microbiol. Rev., 46, 43–70.

Körner, R., B. and Ottow, J. C. G., 1993, 'Quantifieierung der Lachgas (N_2O) – Freisetzung aus Klaranlagen unterschiedlicher Verfahrensführung. Quantification of the release of laughing gas (N_2O) from sewage treatment works with different process control systems, Korrespondenz Abwasser; 40 (4), 514, 516, 518, 520, 522, 524–525.

Lemaire, R., Meyer, R., Taske, A., Crocetti, G. R., Keller, J. and Yuan, Z., 2006, Identifying causes for N_2O accumulation in a lab-scale sequencing batch reactor performing simultaneous nitrification, denitrification and phosphorus removal. Journal of Biotechnology 122 (1), 62–72.

Lens, P. N., De Poorter, M.-P., Cronenberg, C. C. and Verstraete, W. H., 1995, Sulfate reducing and methane producing bacteria in aerobic wastewater treatment systems, Water Research (29–3) pp. 871–880

Mohanakrishnan, J., Gutierrez, O., Meyer, R. L. and Yuan, Z. 2008, Nitrite effectively inhibits sulfide and methane generation in a laboratory scale sewer reactor. Water Research, 42 (14), 3961– 3971.

Mohanakrishnan, J., Gutierrez, O., Sharma, K. R., Guisasola, A., Meyer, R., Keller, J. and Yuan, Z. 2009, Impact of nitrate addition on biofilm properties and activities in rising main sewers. Water Research 43 (17), 4225–4237.

Noda, N., Kaneko, N., Mikami, M., Kimochi, Y., Tsuneda, S., Hirata, A., Mizuochi, M. and Inamori, Y., 2003, Effects of SRT and DO on N_2O reductase activity in an anoxic-oxic activated sludge system. Water Science and Technology 48 (11–12), 363–370.

Osada, T., Kuroda, K. and Yonaga, M., 1995, Reducing nitrous oxide gas emissions from fill-and-draw type activated sludge process. Water Research 29 (6), 1607–1608.

Otte, S., Grobben, N. G., Robertson, L. A., Jetten, M. S. M. and Kuenen, J. G., 1996, Nitrous oxide production by *Alcaligenes faecalis* under transient and dynamic aerobic and anaerobic conditions, Applied and Environmental Microbiology, July, pp. 2421–2426.

Otte, S., 2000, Nitrous Oxide (N_2O) production during conversion of inorganic nitrogen compounds in pure and mixed cultures, PhD thesis, TU Delft.

Park, K. Y., Inamori, Y., Mizuochi, M. and Ahn, K. H., 2000, Emission and control of nitrous oxide from a biological wastewater treatment system with intermittent aeration. Journal of Bioscience and Bioengineering 90 (3), 247–252.

Riet, van't K., 1979, Review of Measuring Methods and Results in Nonviscous Gas-Liquid Mass Transfer in Stirred Vessels. Industrial and Engineering Chemistry Process Design and Development 18 (3), 357–364.

Ritchie, G. A. F. and Nicholas, D. J. D., 1972, Identification of the sources of nitrous oxide produced by oxidative and reductive processes in Nitrosomonas europaea. Journal Biochemistry., 126, 1181–1191.

Schalk-Otte, S., Seviour, R. J., Kuenen, J. G. and Jetten, M. S. M., 2000, Nitrous oxide (N_2O) production by Alcaligenes faecalis during feast and famine regimes. Water Research 34 (7), 2080–2088

Schulthess, R. V., Wild, D. and Gujer, W., 1994, Nitric and nitrous oxides from denitrifying activated sludge at low oxygen concentration. Water Science and Technology 30 (6), 123–132.

Schulthess, R.v., Kuehni, M. and Gujer, W. 1995, Release of nitric and nitrous oxides from denitrifying activated sludge. Water Research 29 (1), 215–226.

Schulthess, R. V., Kuehni, M. and Gujer, W., 1995, Release of nitric and nitrous oxides from denitrifying activated sludge. Water Research 29 (1), 215–226.

Sharma, K. R., Yuan, Z., de Haas, D., Hamilton, G., Corrie, S. and Keller, J. 2008, Dynamics and dynamic modelling of H2S production in sewer systems. Water Research 42 (10–11), 2527–2538.

Sommer, J., Ciplak, A., Sümer, E., Benckiser, G. and Ottow, J. C. G. (1998) Quantification of emitted and retained N_2O in a municipal wastewater treatment plant with activated sludge and nitrification denitrification units. Agrobiological Research 51 (1), 59–73.

Stenstrom, M. K., 2007. Measurement of Oxygen Transfer in Clean Water, ASCE/EWRI 2–06. American Society of Civil Engineers

Stuven, R., Vollmer, M. and Bock, E., 1992, The impact of organic matter on nitric oxide formation by *Nitrosomonas europaea*. Archives Microbiology., 158, 439–443.

Sudarjanto, G., Sharma, K. R., Gutierrez, O. and Yuan, Z. 2010a, A laboratory assessment of the impact of brewery wastewater discharge on sulfide and methane production in a sewer. Proceedings of the 6th International Conference on Sewer Processes and Networks, Nov. 7–11, Gold Coast, Australia.

Sudarjanto, G., Sharma, K. R., Gutierrez, O. and Yuan, Z. 2010b, A laboratory assessment of the impact of dairy wastewater discharge on sulfide and methane production in a sewer. In preparation.

Sümer, E., Weiske, A., Benckiser, G. and Ottow, J. C. G., 1995, Influence of environmental conditions on the amount of N_2O released from activated sludge in a domestic waste water treatment plant. Cellular Molecular Life Science, 51, 419–422.

Tallec, G., Garnier, J., Billen, G. and Gousailles, M., 2006a, Nitrous oxide emissions from secondary activated sludge in nitrifying conditions of urban wastewater treatment plants: effect of oxygenation level. Water Research 40 (15), 2972–2980.

Tallec, G., Garnier, J. and Gousailles, M., 2006b, Nitrogen removal in a wastewater treatment plant through biofilters: nitrous oxide emissions during nitrification and denitrification. Bioprocess and Biosystems Engineering 29 (5–6), 323–333.

Tallec, G., Garnier, J., Billen, G. and Gousailles, M., 2008, Nitrous oxide emissions from denitrifying activated sludge of urban wastewater treatment plants, under anoxia and low oxygenation. Bioresource Technology 99 (7), 2200–2209.

Tamimi, A., Rinker, E. B. and Sandall, O. C., 1994. Diffusion Coefficients for Hydrogen Sulfide, Carbon Dioxide and Nitrous Oxide in Water over the Temperature Range 293–368 K. Journal of Chemical and Engineering Data 39 (2), 330–332.

Tchobanoglous, G., Burton, F. L. and Stensel, H. D., 2003. Wastewater Engineering, Treatment and Reuse, 4th edition. McGraw Hill, Boston.

Thoern, M. and Soerensson, F., 1996, Variation of nitrous oxide formation in the denitrification basin in a wastewater treatment plant with nitrogen removal. Water Research 30 (6), 1543–1547.

Tsuneda, S., Mikami, M., Kimochi, Y. and Hirata, A., 2005, Effect of salinity on nitrous oxide emission in the biological nitrogen removal process for industrial wastewater. Journal of Hazardous Materials 119 (1–3), 93–98.

UNFCCC, 2007, Uniting on Climate. United Nations Framework Convention on Climate Change, Bonn.

Usepa, 2009, Inventory of U. S. Greenhouse Gas Emissions and Sinks: 1990–2006, EPA 430-R-08–005. Washington, D. C.

Vanrolleghem, P. A., Sin, G. and Gernaey, K. V., 2004, Transient response of aerobic and anoxic activated sludge activities to sudden substrate concentration changes. Biotechnology and Bioengineering 86 (3), 277–290.

Weiss, R. F. and Price, B. A., 1980, Nitrous oxide solubility in water and seawater. Marine Chemistry 8 (4), 347–359.

Wicht, H. and Beier, M., 1995, N_2O emission aus nitrifizierenden und denitrificierended Klaranlagen. Korrespondenz Abwasser, 42, 404–406, 411–413.

Yu, R., Kampschreur, M. J., Loosdrecht, M. C. M. V. and Chandran, K. 2010, Mechanisms and Specific Directionality of Autotrophic Nitrous Oxide and Nitric Oxide Generation during Transient Anoxia. Environmental Science Technology, 44, 1313–1319.

Zeng, R. J., Yuan, Z. and Keller, J., 2003, Enrichment of denitrifying glycogen-accumulating organisms in anaerobic/anoxic activated sludge system. Biotechnology and Bioengineering 81 (4), 397–404.

Zhang L., Keller, J. and Yuan, Z. 2009, Inhibition of sulfate-reducing and methanogenic activities of anaerobic sewer biofilms by ferric iron dosing. Water Research 43 (17), 4123–4132.

Zheng, H., Hanaki, K. and Matsuo, T., 1994, Production of nitrous oxide gas during nitrification of wasterwater. Water Science and Technology 30 (6), 133–141.

Zhou, Y., Pijuan, M., Zeng, R. J. and Yuan, Z. 2008, Free Nitrous Acid Inhibition on Nitrous Oxide Reduction by a Denitrifying-Enhanced Biological Phosphorus Removal Sludge. Environmental Science Technology, 42, 8260–8265

Zumft, W. G. 1992, The denitrifying prokaryotes. IN Balows, A., H. G. Truper, M. Dworkin, W. Harder and K. H. Schleifer (Ed.) The Prokaryotes. 2 ed. Berlin, Springer-Verlag.

Zumft, W. G.,1997, Cell biology and molecular basis of denitrification. Microbiol.Mol. Biol. Rev., 61, 533–616.

Appendix 2.1

Overview applied emission factors CH_4

<table>
<tr><th>Country</th><th>Details of methodology</th><th>Ref. in NIR</th></tr>
<tr><td>UK</td><td>Estimates based on Hobson et al. (1996) who estimated emissions for 1990-1995. Subsequent years extrapolated on the basis of population. Empirical EFs have been calculated for 14 different treatment routes. The method complies with the IPCC Good Practice Guidance as a national model (IPCC, 2000).</td><td>Hobson et al. (1996)</td></tr>
<tr><td>Australia</td><td>IPCC default methodology. Tier 2 methodology developed for industrial wastewater.</td><td></td></tr>
<tr><td>Canada</td><td>Based on IPCC methodology; uses EF of 4.015 kgCH_4/person/year for CH_4 emissions from wastewater sector. No evaluation of emissions for industrial wastewater due to lack of activity data.</td><td>ORTECH Corp. (1994)</td></tr>
<tr><td>Denmark</td><td>IPCC Tier 2 country-specific methodology. According to IPCC GL (1996), domestic and industrial wastewaters (and respective sludge) should be accounted separately. In Denmark, most industrial wastewater is treated at centralized WWTWs and the data available for the Total Organics in Wastewater (TOW) does not differentiate between industrial and municipal sewage sludge. Therefore it was assumed that the characteristics of TOW in domestic and industrial wastewater are similar and the division between the two types of sewage sludge was derived by subtracting the %age industrial contribution to the total emissions. The IPCC default methodology was applied from 1990 to 1998, by accounting and correcting for the industrial influent load. Since 1999 onwards actual data from the industry sector has been collected and the TOW calculated using country-specific data. The EF is calculated by the IPCC methodology (1996) using the default value of 0.6 kg CH_4/kg BOD but country-specific values for the MCF. The following constant values have been extrapolated: 0.26 kg CH_4/kg BOD based on the dry weight fraction of anaerobically treated sludge and 0.15 kg CH_4/kg BOD based on the wet weight fraction. The fractions that are used for biogas, combustion or reuse including combustion include methane potentials that are recovered or emitted as CO_2. The estimated CH_4 potentials of these fractions have been estimated and subtracted from the calculated gross emission of CH_4.</td><td>Thomsen and Lick (2005)</td></tr>
<tr><td>Germany</td><td>Municipal wastewater treatment uses aerobic procedures, therefore it produces no CH_4 emissions (Methane Correction Factor: MCF=0).CH_4 emissions from cesspools and septic tanks (which is exceptional in Germany) are determined following the IPCC methodology.</td><td>FEA (2008)</td></tr>
<tr><td>France</td><td>Minor variation on IPCC default methodology.</td><td></td></tr>
<tr><td>Japan</td><td>Country specific EFs were developed:
1) From sewage treatment plants: 528.7 mg CH_4/m^3 (to secondary treatment) & 348.0 mg CH_4/m^3 (from sludge); uncertainty in emissions 33% (95%ile).
2) From in-situ domestic sewage treatment (septic tanks):
<table>
<tr><th>Domestic wastewater treatment facilities</th><th>CH_4 EF (kg CH_4/person-year)</th></tr>
<tr><td>Community plants</td><td>0.195</td></tr>
<tr><td>Gappei-shori johkasou</td><td>1.106</td></tr>
<tr><td>Tandoku-shori johkasou</td><td>0.197</td></tr>
<tr><td>Vault toilets</td><td>0.197</td></tr>
</table>
87% uncertainty in emissions estimation (95%ile)
3) From collected domestic sewage (septic tanks)
<table>
<tr><th>Treatment method</th><th>CH_4 EF (kg CH_4/m^3)</th></tr>
<tr><td>Anaerobic treatment</td><td>0.543</td></tr>
<tr><td>Aerobic treatment</td><td>0.00545</td></tr>
<tr><td>Standard de-nitrification treatment</td><td>0.0059</td></tr>
<tr><td>High load de-nitrification treatment</td><td>0.005</td></tr>
</table></td><td>GIO (2008)</td></tr>
</table>

Country	Details of methodology	Ref. in NIR
	Membrane separation: 0.00545 Other: 0.00545 101% uncertainty in emissions estimation (95%ile)	
Netherlands	IPCC Tier 2 method. Anaerobic industrial wastewater treatment: EF = 0.056 t CH_4/t DOC design capacity. Urban wastewater treatment (and anaerobic sludge handling): EF = 0.085 t CH_4/t DOC influent. Septic tanks: EF = 0.0075 t CH_4/year/person (MCF=0.5; B_0=0.25)	
New Zealand	Refinement of the IPCC methodology (1996). Characteristics EFs have been calculated for 9 typical on-site treatment processes (largely used in the Country). They have been calculated from the proportion of BOD to the plant which is anaerobically degraded, multiplied by the CH_4 conversion factor MCF. IPCC default methodology also used for industrial wastewater.	SCS Wetherhill Environmental, (2002); Beca (2007)
Sweden	Not estimated (either recovered as heat and bio-energy, or insignificant).	
USA	A slightly modified IPCC 1996 methodology was used. CH_4 emissions from septic systems, WWTWs and sludge digestion are calculated separately. The IPCC default factors (B_0 and MCF) are used. Emissions from industrial wastewater were developed according to the IPCC (1996) methodology (four big industries producing methane have been identified and activity data collected)	USEPA (2008)
Austria	IPCC method with MCF for septic tanks = 0.27 (calculated from average temperatures in Austria: 20°C for 2/3 of the year with MCF = 35% and 10°C for 1/3 of the year with a MCF = 10%)	Steinlechner *et al.* (1994)

Appendix 2.2

Dimension WWTPs and sampling points USA

Legend

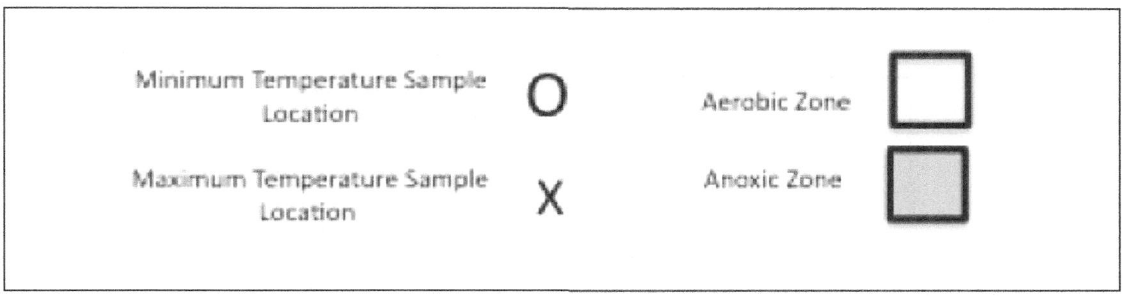

Separate-stage BNR

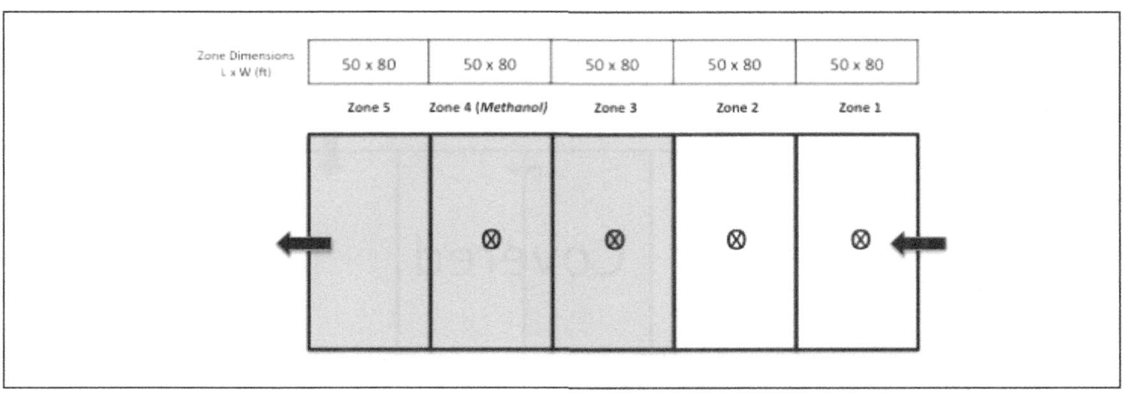

Four-stage Bardenpho

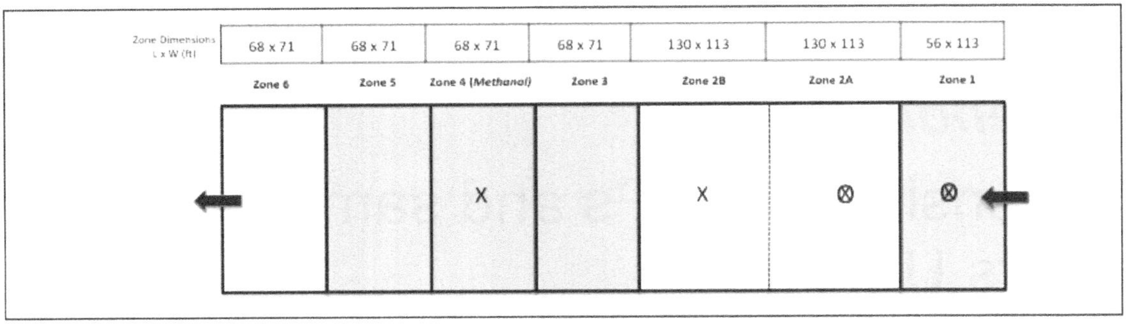

Step-feed BNR 1

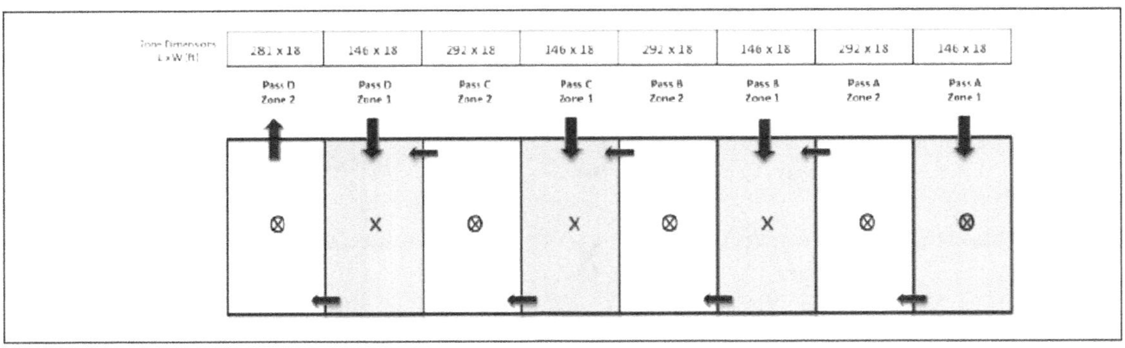

Step-feed non BNR

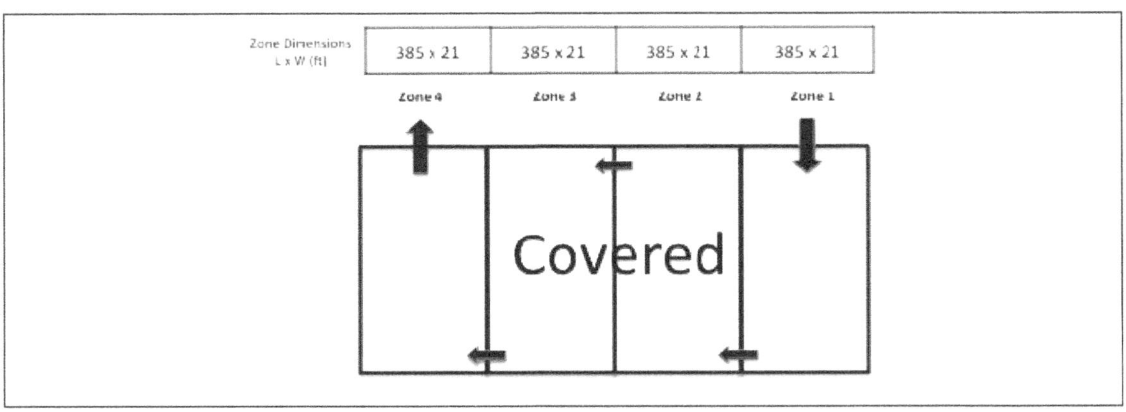

Separate centrate

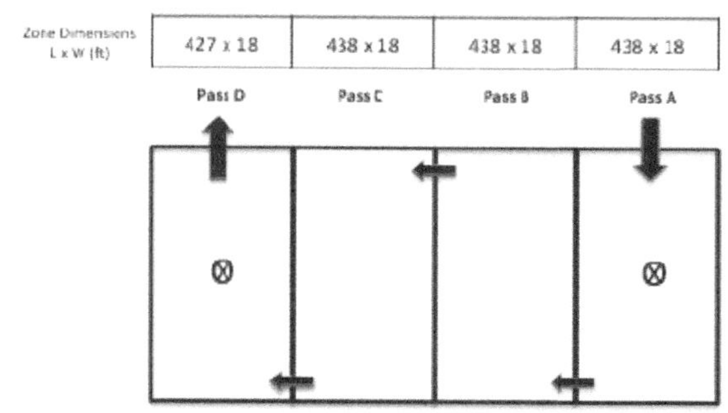

Plug-flow 1

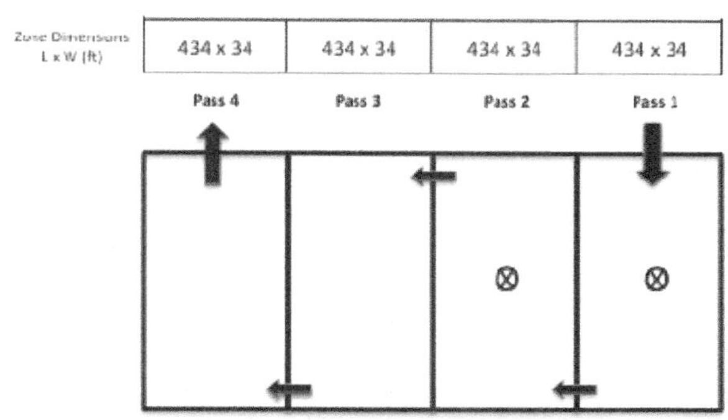

Plug-flow 2

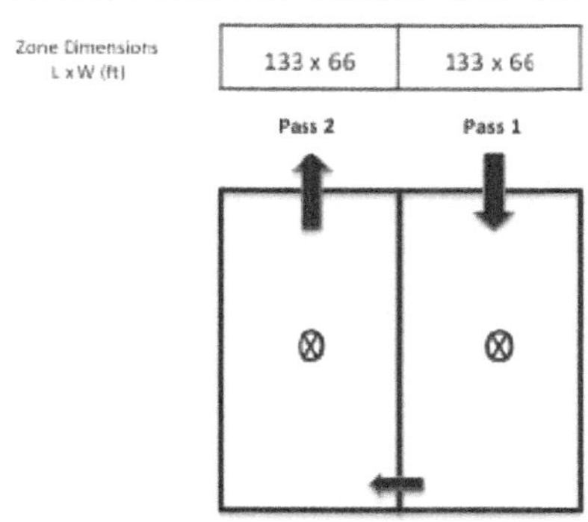

MLE 1

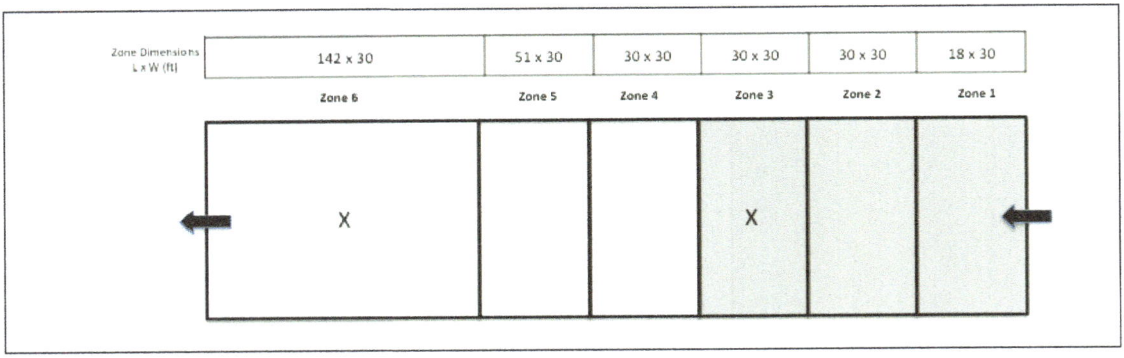

MLE 2

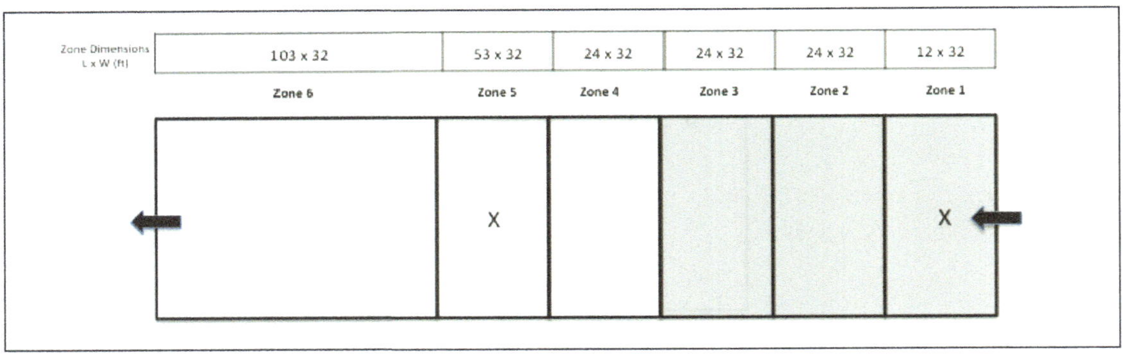

Step-feed BNR 2

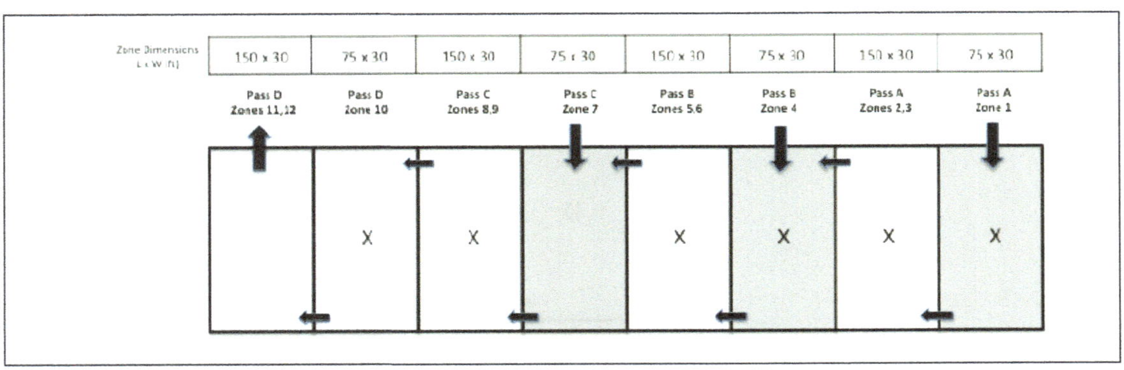

Oxidation ditch

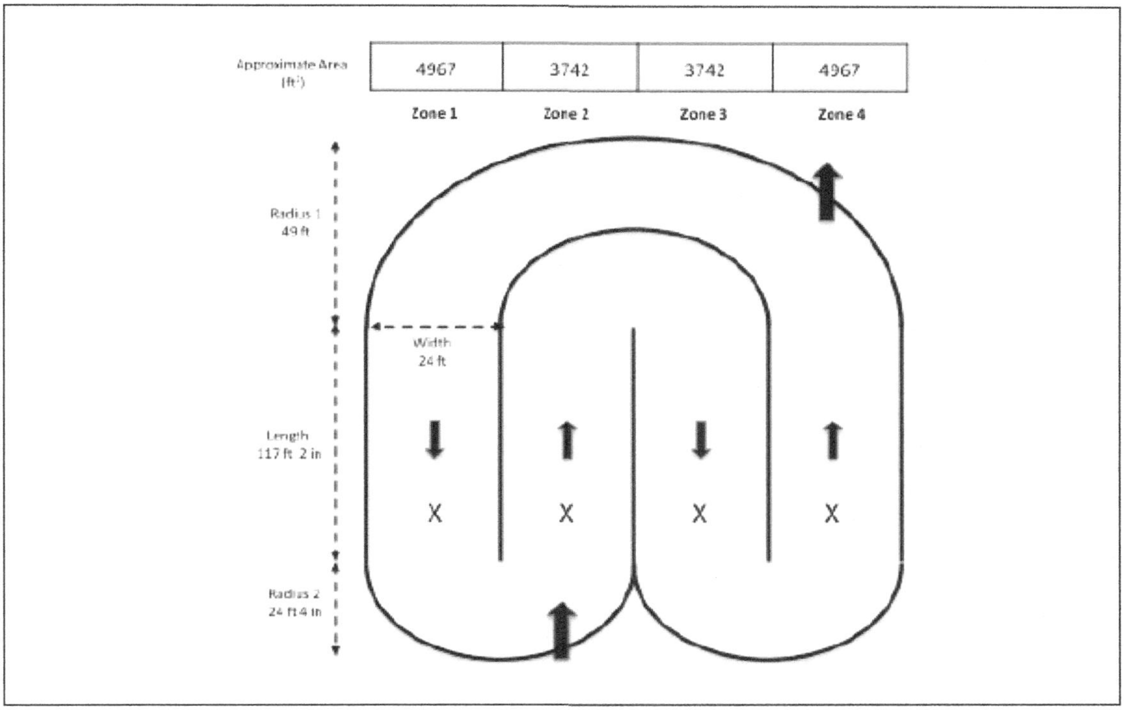

Step-feed BNR 3

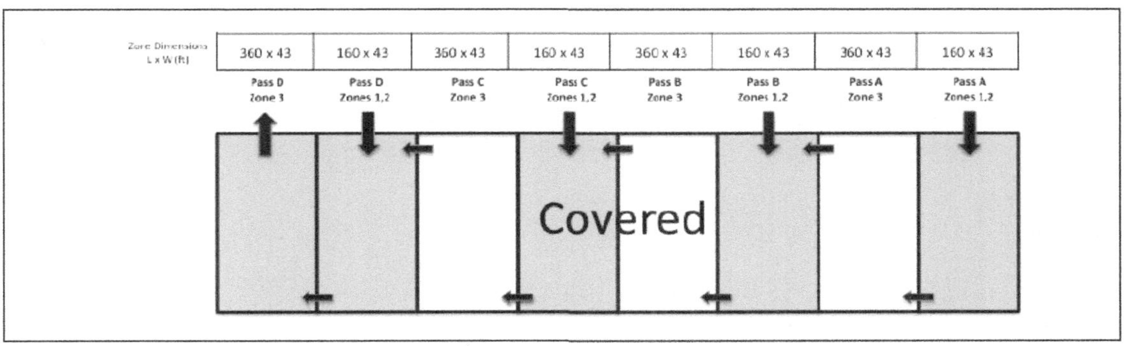

Appendix 2.3
Locations of N_2O and CH_4 measurements in the Netherlands

WWTP Papendrecht

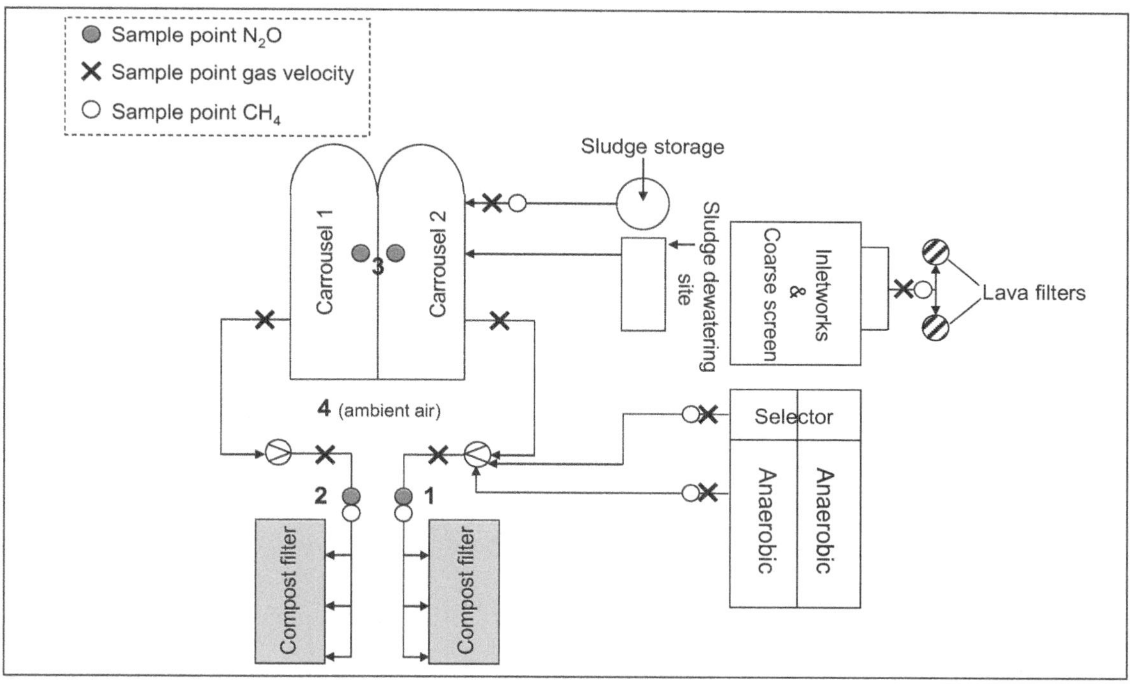

WWTP Kortenoord

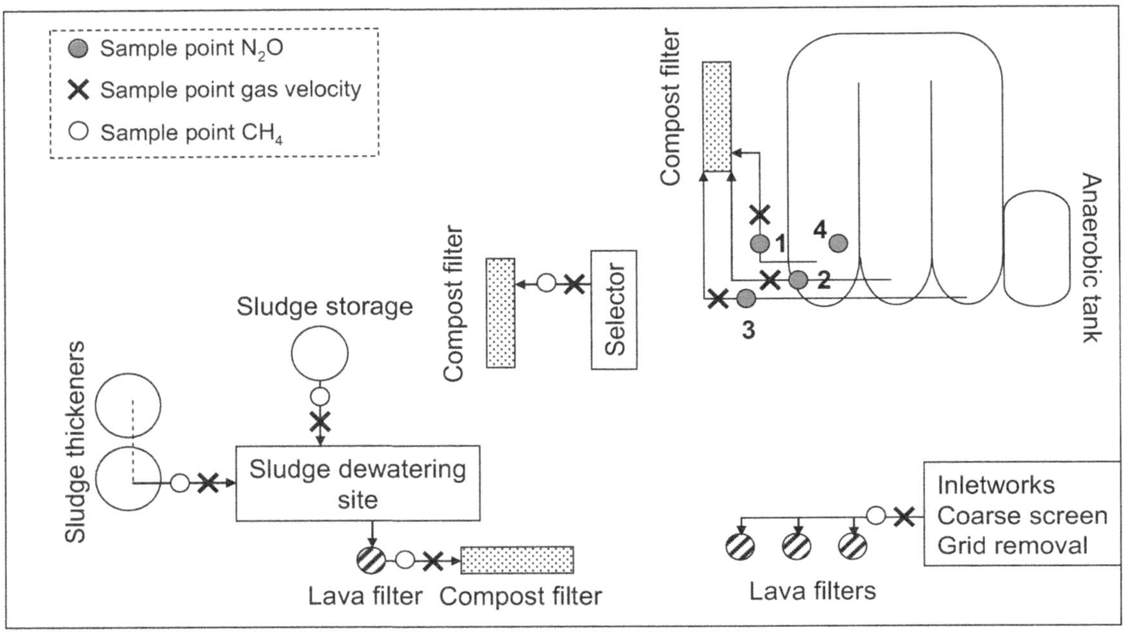

WWTP Kralingseveer

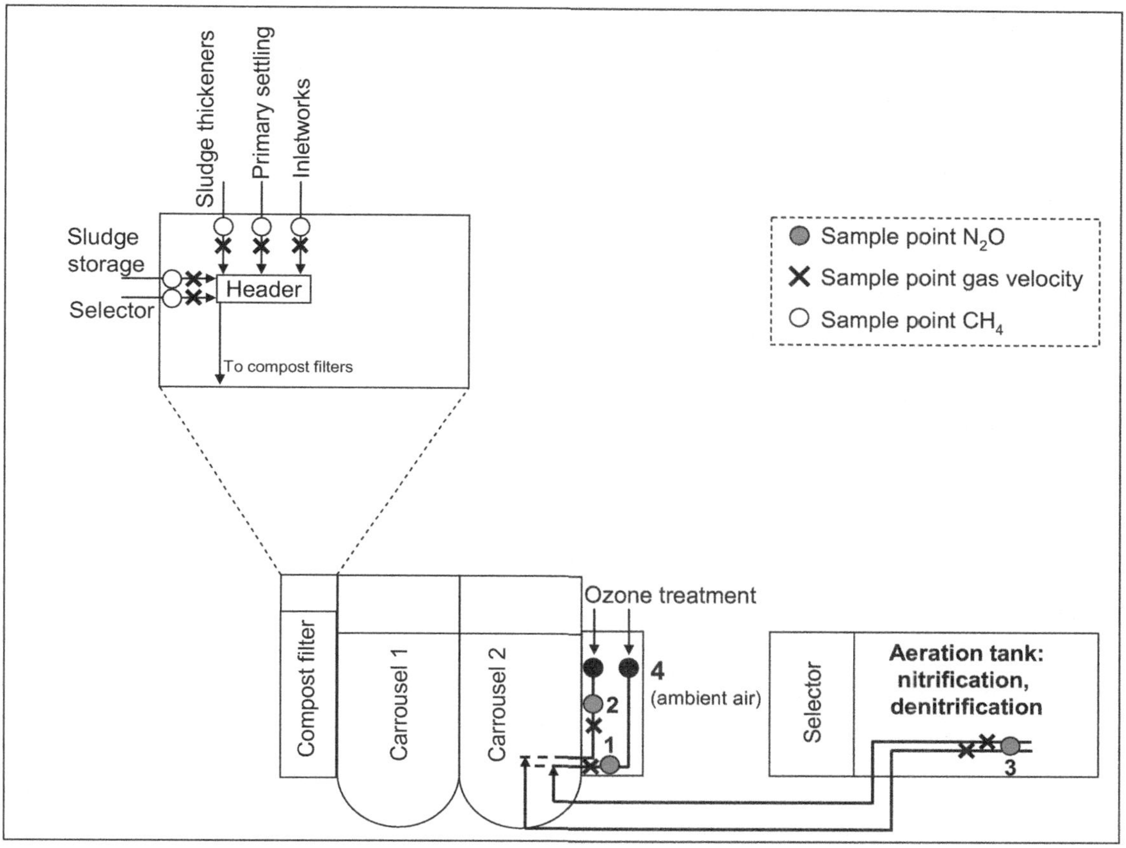

Appendix 2.4
Abbreviations and Glossary

Abbreviation/term	
A/O and A^2/O	Type of WWTP configuration for the biological removal of phosphate[1]
ADEME	French Environment and Energy Management Agency
Bardenpho	Type of WWTP configuration for the biological removal of phosphate[1]. Can be designed with different amount of stages.
BNR	Biological Nitrogen Removal
CH_4	Methane
CO_2	Carbon Dioxide
COD	Chemical Oxygen Demand
DO	Dissolved Oxygen
EPA	Environmental Protection Agency
FB	Fermentative Bacteria
$Fe^{2/3+}$	Iron
FID	Flame Ionization Detector
GC	Gas Chromatograph
GF_{WWTP}	Generation factor (N_2O) for whole WWTP
GHG	Greenhouse Gas
H_2S	Hydrogen Sulphide
HRT	Hydraulic Retention Time
IPCC	Intergovernmental Panel on Climate Change
Johannesburg	Type of WWTP configuration[1]
K_La	Mass transfer coefficient (liquid → gas)

(*Continued*)

Abbreviation/term	
LCC - PS	Lower Crooked Creek Pump Station
MBR	Membrane Bioreactor
MLE	Modified Ludzack-Ettinger (i.e. a configuration type)
NH_4^+ (-N)	Ammonium
NGERS	National Greenhouse and Energy Reporting System
NO_2^- (-N)	Nitrite
$^-$ (-N)	Nitrate
N_2O	Nitrous Oxide
POTW	Public Owned wastewater treatment
SBR	Sequenced Batch Reactor
SRB	Sulphate Reducing Bacteria
SRT	Sludge retention time
STOWA	Foundation for Applied Water Research
TKN	Total Kjeldahl Nitrogen
Ton (USA)	US ton is 2,000 pounds (907 kg)
USEPA	United States Environmental Protection Agency
UKWIR	UK Water Industry Research
VFA	Volatile Fatty Acids
WERF	Water Environment Research Foundation
WSAA	Water Service Assocation of Australia
WWTP	Wastewater Treatment Plant

[1] For more information see: Tchobanoglous, G., 2003

Appendix 2.5

National greenhouse and energy reporting system, Australia

Facility Definition

Of particular relevance to water industry is the definition of a 'facility', with respect to water supply, sewerage or drainage services. The provision from the NGER Regulations (see the box below) would treat all of the activities undertaken as part of a water supply, sewerage and drainage services network as a single undertaking or enterprise and therefore one facility for the purposes of the Act's threshold.

In 2009, the Water Services Association of Australia (WSAA) requested a determination from the DCCEE on an alternative definition of the boundaries of 'facilities' for the water supply, sewerage and drainage services sector. This submission, at a minimum, sought to separate 'water facilities' into two separate continuous systems:

1. From water source and extraction, through treatment and distribution; and
2. Sewage collection, sewage treatment and effluent discharge.

These sub-systems would then be further disaggregated, based on separate physical catchments or distribution zones. Allocation of emissions to different corporate entities will then depend on who has 'operational control' of the various elements of the 'facility'. Presently, the DCCEE has advised that they will not make a formal response to WSAA's request.

2.20 *Forming part of a single undertaking or enterprise — electricity, gas etc activities*

(1) *If:*

(a) *an activity in a series of activities is in one of the industry sectors mentioned in sub-regulation (2); and*

(b) *the activity and any ancillary activities to it are under the overall control of the same corporation;*

then all of the activities will form part of the same single undertaking or enterprise.

(2) *For sub-regulation (1), the industry sectors are as follows:*

(a) *Electricity transmission (262);*

(b) *Electricity distribution (263);*

(c) *Gas supply (270);*

(d) *Water supply, sewerage and drainage services (281);*

(e) *Telecommunications services (580)*

Choice of Calculation Method

The NGERS *Technical Guidelines 2009* provide Methods that broadly allow for both direct emissions monitoring and the estimation of emissions through the tracking of observable, closely-related variables. Four Methods have been described which provide a framework for emissions estimation for a range of purposes:

- **Method 1**: the National Greenhouse Accounts default method:
  - Method 1 provides a class of estimation procedures derived directly from the methodologies used by the Department of Climate Change for the preparation of the *National Greenhouse Accounts*[1]. This is the most basic of the available methods.
- **Method 2**: a facility-specific method using industry sampling and Australian or international standards listed in the Determination to provide more accurate estimates of emissions at the facility level:
  - Method 2 enables corporations to undertake additional measurements in order to gain more accurate estimates for emissions for that particular facility.
- **Method 3**: a facility-specific method using Australian or international standards listed in the Determination or equivalent standards for both sampling and analysis:
  - Method 3 is very similar to Method 2, except that it requires, additionally, reporters to comply with Australian or international documentary standards for sampling, as well as documentary standards for the analyses.
- **Method 4**: direct monitoring of emission systems, either on a continuous or periodic basis:
  - Method 4 provides approaches to enable direct monitoring of greenhouse gas emissions arising from an activity. This can provide a higher level of accuracy in certain circumstances, depending on the type of emission process although it is more data intensive than other approaches.
  - There is presently no Method 4 defined for wastewater handling emissions.

---

[1]This involves estimation of COD and nitrogen loads into the plant based on the following default factors: A default quantity of COD per capita of 0.0585 tonnes per person per year; and a default nitrogen load of 0.16 x 0.036 = 0.00576 tonnes per person per year.

The provision to select a Method for the estimation of emissions allows organisations to make their own judgement on the balance of costs of using the higher-order methods, with the benefits of potentially improved emission estimates. Experience suggests that the use of Method 2 for estimating emissions related to wastewater treatment generally results in approximately 15% lower results when compared to results calculated using Method 1. Therefore, where sufficient activity data is readily available, it is generally advisable to use Method 2, as it is less likely to over-estimate emissions than Method 1.

Methodology for Estimating CH_4 Emissions

Presently, there is no NGERS methodology to estimate fugitive greenhouse gas emissions arising from methane generation in the sewerage network, upstream of wastewater treatment facilities. The NGERS *Technical Guidelines 2009* promulgates the following equations for the estimation of fugitive CH_4 emissions arising from wastewater treatment and handling:

Equation 1
(NGERS Technical Guidelines, Division 5.2.3 Section 5.25 Sub-section (1)):

$$E_j = CH_4^* - \gamma(Q_{cap} + Q_{flared} + Q_{tr})$$

where:

- Ej is the emissions of CH_4 released by the plant during the year, measured in CO_2-e tonnes;
- CH_4^* is the estimated quantity of CH_4 in biogas released by the plant during the year, measured in CO_2-e tonnes;
- δ is the factor 6.784 × 10-4 × 21 converting cubic metres of CH_4 at standard conditions (15°C, 1 atm) to CO_2-e tonnes;
- Q_{cap} is the quantity of CH_4 in biogas captured for combustion for use by the plant during the year, measured in cubic metres;
- Q_{flared} is the quantity of CH_4 in biogas flared during the year by the plant, measured in cubic metres; and
- Q_{tr} is the quantity of CH_4 in biogas transferred out of the plant during the year, measured in cubic metres.

The intent of Equation 1 is to reconcile the estimated **theoretical** amount of CH_4 generated at the wastewater treatment plant (CH_4^*) with the actual amount of CH_4 **measured** in the captured biogas (Q_{cap}, Q_{flared} and Q_{tr}). Any difference between these two figures, and assuming $CH_4^* > \gamma(Q_{cap} + Q_{flared} + Q_{tr})$, is hence assumed to equal the amount of 'fugitive' CH_4 emitted to the atmosphere.

Equation 2
(NGERS Technical Guidelines, Division 5.2.3 Section 5.25 Sub-section (5)):

$$CH_4^* = (COD_w - COD_{sl} - COD_{eff}) \times F_{wan} \times EF_{wij} + (COD_{sl} - COD_{trl} - COD_{tro}) \times F_{slan} \times EF_{slij}$$

where:

- COD_w is the quantity of COD in wastewater entering the plant during the year, measured in tonnes;

- COD_{sl} is the quantity of COD removed as sludge from wastewater and treated in the plant, measured in tonnes of COD;
- COD_{eff} is the quantity of COD in effluent leaving the plant during the year, measured in tonnes;
- F_{wan} is the fraction of COD anaerobically treated by the plant during the year;
- EF_{wij} is the default CH_4 emission factor for wastewater with a value of 5.3 CO_2-e tonnes per tonne COD;
- COD_{trl} is the quantity of COD in sludge transferred out of the plant and removed to landfill, measured in tonnes of COD;
- COD_{tro} is the quantity of COD in sludge transferred out of the plant and removed to a site other than landfill, measured in tonnes of COD;
- F_{slan} is the fraction of COD in sludge anaerobically treated by the plant during the year.
- EF_{slij} is the default CH_4 emission factor for sludge with a value of 5.3 CO_2-e tonnes per tonne COD (sludge).

A simple process flow diagram is included in Figure A2.5.1 to help clarify the definition of terms in Equation 2.

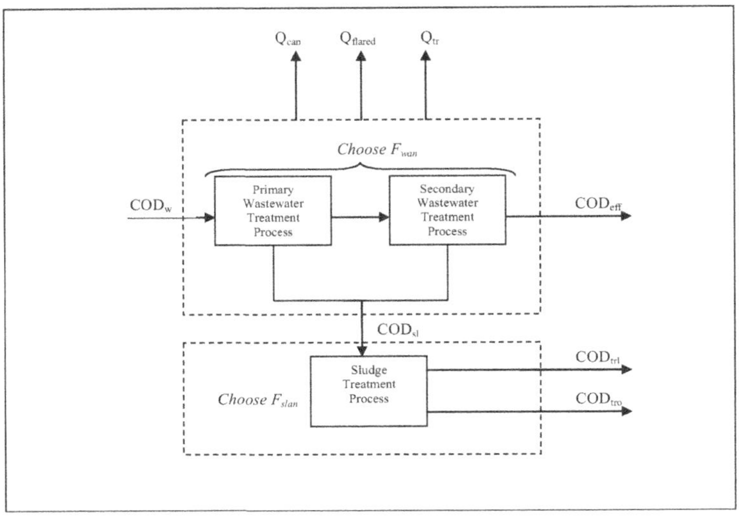

Figure A2.5.1 Default NGERS COD Mass Balance.

The intent of Equation 2 is to estimate the theoretical quantity of CH_4 generated at the plant, based on a simple COD mass balance. The equation is divided into two parts. The first part is a simple COD mass balance over the main wastewater treatment processes (i.e. 'liquid train' primary and secondary treatment). The second part is a simple COD mass balance over the sludge treatment processes (e.g. anaerobic digesters, sludge lagoons, dewatering, etc.).

Each COD mass balance determines how much COD is consumed in the treatment process by converting it to a gas (i.e. CO_2 or CH_4). Multiplying by the maximum CH_4 generation factor (EF) calculates how much CH_4 would be generated if **100%** of the COD consumed was converted to CH_4. Then, multiplying by the process-specific F_{wan} or F_{slan} factor calculates the expected fraction of CH_4 emissions for that particular type of process.

Clarification on fwan and fslan factors

The NGERS definition of F_{wan} as 'the fraction of COD anaerobically treated by the plant during the year' and of F_{slan} as 'the fraction of COD in sludge anaerobically treated by the plant during the year' can be misleading and confusing.

To be consistent with Intergovernmental Panel on Climate Change (IPCC) terminology and to provide a better description of its true meaning, F_{wan} can be re-defined as:

> 'MCFw – the CH_4 correction factor for wastewater treatment, or fraction of the maximum CH_4 emission factor'.

The NGERS definition of EFwij can also be redefined, to be consistent with IPCC terminology, as 'the maximum CH_4 generation factor for wastewater'. This value is defined by the chemical stoichiometry of CH_4 production. In practice however, this maximum is not achieved (even in 100% anaerobic systems) due to some (or all) of the following losses:

- Some COD may be consumed via other biochemical pathways (e.g. sulphide reduction, volatile acids production);
- Some CH_4 that is generated may be consumed by aerobic methanotrophic microorganisms and converted to carbon dioxide (i.e. through surface layer oxidation in uncovered lagoons); and
- Some CH_4 that is generated may remain dissolved in the liquid phase (NOTE: dissolved CH_4 is not accounted for in standard COD analysis[2]).

These losses are accounted for by the F_{wan} factor. For example, the default NGERS F_{wan} value of 0.8 for anaerobic deep lagoons accounts for the typical CH_4 losses in open anaerobic lagoon systems. This equates to 80% of consumed COD being converted to gaseous CH_4 as an end product, with the balance of 20% consumed via other pathways or remaining in its soluble form (and therefore not adding to overall atmospheric CH_4 emissions).

Similarly, to be consistent with IPCC terminology and to provide a better description its true meaning, F_{slan} can also be redefined as:

> 'MCFsl – the CH_4 correction factor for sludge treatment, or fraction of the maximum CH_4 emission factor'.

EF_{slij} can also be redefined as 'the maximum CH_4 generation factor for sludge'.

Based on verbal discussions with the DCCEE, the formal definitions of F_{wan} and F_{slan} are likely to change, in line with these recommendations, in the next major revision of the NGERS *Determination* and associated *Technical Guidelines* in 2011.

The NGERS *Technical Guidelines* include IPCC default F_{wan} and F_{slan} factors (equivalent to IPCC MCFs) for various types of treatment systems. These factors have been compiled by the IPCC and are based on 'expert judgement by lead authors of this section'[3].

The IPCC Guidelines also include ranges for each MCF to account for the variation that is likely to be seen in practice for individual treatment systems. These ranges (shown in Table A2.5.1 below) are not promulgated under NGERS.

---

[2]Hartley, K., Lant, P. (2006). Eliminating non-renewable CO_2 emissions from sewage treatrment: An anaerobic migrating bed reactor pilot plant study, *Biotechnology and Bioengineering*, 95(3): 384-398.

[3]IPCC, (2006). *2006 IPCC Guidelines for National Greenhouse Gas Inventories*, Ch.6 – Wastewater Treatment and Discharge, Table 6.3 p. 6.13.

Table A2.5.1 NGERS Default Correction Factors and IPCC Ranges for Some Treatment Systems.

Treatment System	F_{wan} or F_{slan} (NGERS)	IPCC Range
Managed Aerobic Treatment	0	0–0.1
Unmanaged Aerobic Treatment	0.3	0.2–0.4
Anaerobic Digester / Reactor	0.8	0.8–1.0
Anaerobic Shallow Lagoon (< 2 m deep)	0.2	0–0.3
Anaerobic Deep Lagoon (> 2 m deep)	0.8	0.8–1.0

The general approach to calculating N2O emissions from wastewater treatment processes, under the NGERS Technical Guidelines 2009 is outlined below in Table A2.5.2.

Table A2.5.2 NGERS Methodology for Estimating Nitrous Oxide Emissions.

NGERS *Technical Guidelines 2009* reference	**Part 5.3, Division 5.3.6 – Method 2**
Simplified generic mass balance diagram	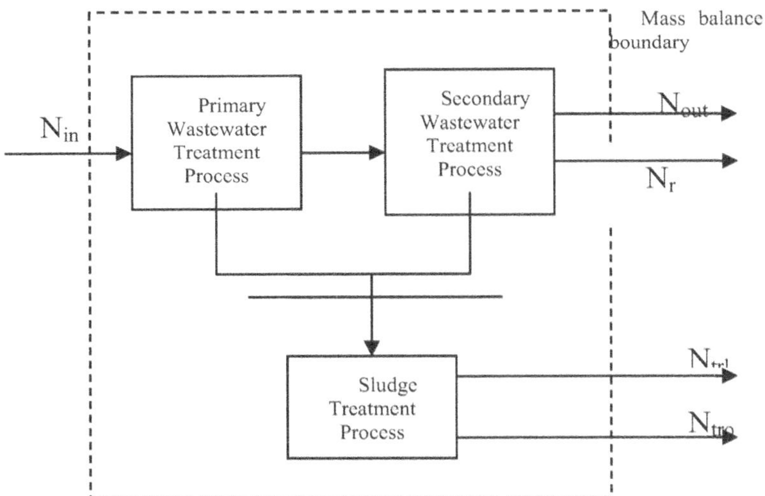
Typical activity data required (units)	• Raw wastewater volume, Q (ML) • Effluent reused by third parties, Q_r (ML) • Mass of biosolids sent to landfill, M_{trl} and/or 'other', M_{tro} (e.g. land application) (tonnes) • Influent total nitrogen concentration, $[N]_{in}$ (mgN/L) • Effluent total nitrogen concentration, $[N]_{out}$ (mgN/L) • Effluent reused by third parties total nitrogen concentration, $[N]_r$ (mgN/L) • Nitrogen content of biosolids, $[N]_b$ (mgN/kg biosolids)

Emission factors, exclusions and assumptions	• Technical Guidelines 2009, Division 5.3.5, section 5.31 (3):
	  ◦ Emission factor for secondary treatment, EF_{secij} = 4.9 tCO_2-e per tonne N removed by denitrification [4];
	  ◦ Emission factor for nitrogen discharge differentiated by the discharge environment, EF_{disij} = 4.9 tCO_2-e per tonne N discharged [5];
	• **One emission factor, EFsecij, applies for ALL processes within a treatment plant, and for ALL types of treatment plants**
	• **One emission factor, EFdisij, applies for ALL types of nitrogen discharges**

Emissions calculation method	• Raw wastewater total N mass load, Nin (tonnes) $= Q \times [N]_{in} \times (10^{-3}\ t/kg)$ • Discharged effluent total N mass load, Nout (tonnes) $= (Q - Q_r) \times [N]_{out} \times (10^{-3}\ t/kg)$ • Total N mass load in water reuse to third parties, Nr (tonnes) $= Q_r \times [N]_r \times (10^{-3}\ t/kg)$ or $Q_r \times [N]_{out} \times (10^{-3}\ t/kg)$ • N mass load to landfill, Ntrl (tonnes) $= M_{trl} \times [N]_b \times (10^{-3}\ t/kg)$ • N mass load to 'other' (e.g. land application), Ntro (tonnes) $= M_{tro} \times [N]_b \times (10^{-3}\ t/kg)$ • Nitrogen mass balance, MBNww (tonnes) (i.e. mass of nitrogen removed by denitrification) $= (N_{in} - N_{out} - N_r - N_{trl} - N_{tro})$ • Greenhouse gas emissions (tCO_2-e) $= MBN_{ww} \times EF_{secij} + N_{out} \times EF_{disij}$

This nitrogen mass balance framework for NGERS calculations is theoretically quite robust, but it is presently limited by the use of a single emission factor for EF_{secij} and EF_{disij}, hence lacking the flexibility to account for different levels of N_2O emissions from different processes and/or operating conditions or effluent discharges.

NGERS Uncertainty

Reporting of uncertainty associated with Scope 1 emissions has been voluntary under NGERS for the first two reporting years. However, the NGERS *Technical Guidelines 2009* now provide extensive guidance on uncertainty estimation in Chapter 8. This guidance is based on the *GHG Protocol Guidance on Uncertainty Assessment in GHG Inventories and Calculating Statistical Parameter Uncertainty*, 2003.

Figure A2.5.2 below shows the types of uncertainty that are associated with the estimation of greenhouse gas emissions. For the purposes of NGERS reporting, it is only necessary to calculate

---

[4]Derivation of EF_{secij}: 10 kg N_2O-N per tonne N removed × (44 kg N_2O per 28 kg N) × (10^{-3} tonnes / kg) × (310 tonne CO_2-e per tonne N_2O) = 4.9 tonne CO_2-e per tonne N removed

[5]Derivation of EF_{disij}: 10 kg N_2O-N per tonne N discharged × (44 kg N_2O per 28 kg N) × (10^{-3} tonnes / kg) × (310 tonne CO_2-e per tonne N_2O) = 4.9 tonne CO_2-e per tonne N discharged

statistical uncertainty – that is, the uncertainty due to the underlying random variability of activity data. It is not necessary to account for systematic uncertainty, such as instrumental error, sampling error or analytical error. Furthermore, it is not necessary to account for the model uncertainty and scientific uncertainty that might be associated with the default emission factors. Hence, the uncertainty calculated for NGERS reporting purposes only represents a fraction of the true uncertainty associated with the results.

Wastewater treatment processes are inherently variable, both in terms of throughput quantities (i.e. flow rates) and qualities (i.e. constituent concentrations), as well as biological performance. Hence, just capturing the statistical uncertainty associated with wastewater handling processes is very difficult. Furthermore, the need to reconcile two completely independent data sets (i.e. Equation 1 above) is also statistically challenging, given that they both have high underlying variability.

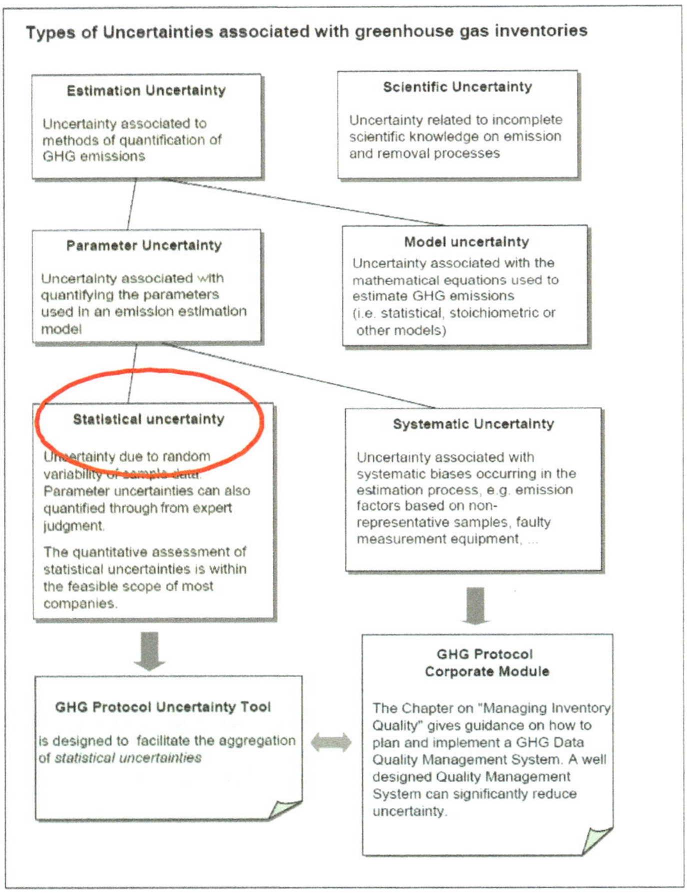

Figure A2.5.2 Types of Uncertainty [(6)]

[6] *GHG Protocol Guidance on Uncertainty Assessment in GHG Inventories and Calculating Statistical Parameter Uncertainty*, 2003

Section 8.10 of the NGERS Technical Guidelines 2009 allows for a default aggregated uncertainty of 40% for domestic wastewater handling. However, by applying the statistical analyses recommended under Chapter 8, it may be possible to refine this uncertainty estimate, based on actual statistical confidence in the activity data (refer section 2.1.8 below).

Calculation Method for NGERS Uncertainty

The NGERS *Technical Guidelines* assumes that each activity data set required for calculations in Equations 1 and 2 is normally distributed. Given this assumption, the statistical uncertainty of each data set can be represented by its arithmetic mean ±95% confidence interval (assuming the Student's *t*-distribution). The 95% percent confidence interval is calculated as follows:

$$\bar{x} \pm t(\tfrac{\alpha}{2}; n-1)\frac{s}{\sqrt{n}}$$

where:

- $\bar{x}$ is the arithmetic mean of the sampling data;
- t is the Student's *t*-distribution;
- α is equal to 0.05 for a 95% confidence interval;
- n is the number of samples in the data set;
- s is the standard deviation of the data set.

The calculations required in Equations 1 and 2 involve addition/subtraction and multiplication/division of these various activity data sets. Combining the statistical uncertainties through these mathematical manipulations is done via a simple first order propagation technique:

- Multiplying/dividing uncertainties:
  - where: (A ± a%) x (B ± b%) = C ± c%,
  - $c = \sqrt{a^2 + b^2}$
- Adding/subtracting uncertainties:
  - where: (C ± c%) + (D ± d%) = E ± e%
  - $e = \frac{\sqrt{(C \times c)^2 + (D \times d)^2}}{E}$

Appendix 2.6
Determination of k_La for N_2O emission, Australia

To determine the N_2O mass transfer coefficient (k_La) for each WWTP reactor zone (refer to Equation 5 in section 14.1.3), a series of lab-scale stripping experiments was conducted. These experiments were conducted in both clean tap water and mixed liquor (from WWTP No.1, MLSS 4.45 $g.L^{-1}$) in a 0.05 m diameter, 0.815 m deep glass column, with a gas sparging bar at its base. Measurements of dissolved N_2O and dissolved oxygen were made at two locations in the liquid column using an on-line Membrane Inlet Mass Spectrometer (Hiden Analytical, Warrington, United Kingdom).

In each stripping experiment, the liquid column was initially saturated by bubbling through 0.51% N_2O gas (in helium), which equated to an equilibrium dissolved N_2O concentration of 6.45 mg.L-1 at 20°C (Weiss and Price, 1980), and dissolved oxygen equal to zero. The sparge gas was then switched to compressed air, and the dissolved concentration of O_2 and N_2O monitored until oxygen reached saturation (9.07 $mg.L^{-1}$ at 20°C, corrected for chlorinity) (APHA, 1995) and N_2O approached zero. This procedure was repeated for clean water and mixed liquor at three different sparge flowrates (500 $mL.min^{-1}$, 200 $mL.min^{-1}$ and 100 $mL.min^{-1}$). In the mixed liquor stripping experiments, the clean water saturation concentrations were reduced by a 95% β correction factor to account for differences in solubility caused by constituents in the mixed liquor such as salts, particulates and surface-active substances (Tchobanoglous et al., 2003).

The N_2O mass transfer coefficient for each experiment was then calculated using a non-linear parameter estimation routine (Stenstrom et al., 1997). The results of these calculations are shown in Figure A2.6.1A. As expected, the mass transfer coefficient decreased with superficial gas velocity, v_g ($m^3.m^{-2}.s^{-1}$), in a power law relationship. This result is similar to other empirical mass transfer modelling approaches for aerated systems, such as bubble columns (Envirosim, 2007; Garcia-Ochoa and Gomez, 2009).

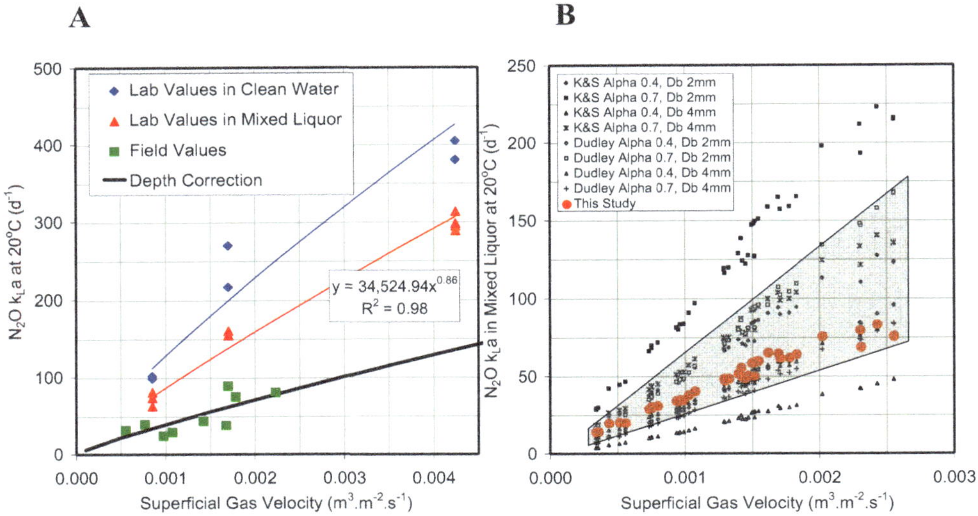

Figure A2.6.1 A) Nitrous oxide volumetric mass transfer coefficients (k_La) from clean water and mixed liquor stripping experiment in a lab-scale column, and from field measurements; B) Comparison of modelled k_La values for all WWTPs (Equation 9) against literature correlations for a range of bubble diameters, D_B and αF factors. The shaded area indicates the 95% confidence interval adopted for the k_La values in Equation 5 in section 5.1.3

However, the volumetric mass transfer coefficient is also related to several other variables, including reactor geometry (particularly aerator immersion depth), aeration bubble size, diffuser layout and liquid viscosity (Gillot et al., 2005). To better correlate the lab-scale bubble column results to full-scale reactors that are significantly deeper (i.e. 3.68 – 5.95 m), six off-gas samples were collected from the aerated zones of WWTP No.5 (depth = 4.5 m), two off-gas samples from WWTP No. 3 (depth = 3.9 m) and one sample from WWTP No. 6 (depth = 4.7 m) (refer to paragraph 5.1.1). For each sample, the mass transfer coefficient was then calculated as per Equation A1:

$$k_L a_F = \frac{\dfrac{Q_A \times p_{N_2O,G}}{R \times T} \times MW_{N_2O}}{V_R \times \left([N_2O]_R - [N_2O]_S^*\right)} \qquad \text{(A1)}$$

Where:

k_La_F = volumetric mass transfer coefficient, calculated from field data (d^{-1})

Q_A = aeration flowrate supplied to reactor at the time of sampling ($m^3.d^{-1}$)

$p_{N2O,G}$ = partial pressure of N_2O in off-gas from reactor (atm)

MW_{N2O} = molecular weight of N_2O (44 $kg.kmol^{-1}$)

R = universal gas constant (0.08206 $m^3.atm.kmol^{-1}.K^{-1}$)

T = reactor liquid temperature at time of sampling (K)

The results for these field k_La estimates are also shown in Figure A2.6.1A, and are lower than the mass transfer coefficients measured for the shallower lab-scale column. Therefore, the power law estimation based on v_g only, from the lab-scale mixed liquor experiments (refer to Figure A2.6.1) was modified by the addition of a correction factor $\left\{\frac{D_R}{D_L}\right\}^{\kappa}$, to account for the increased depth of full-scale WWTP reactors. The value of κ was empirically determined using a sum-of-least-squares fitting algorithm in MS Excel. The resulting depth-corrected kLa correlation is shown in Equation A2:

$$k_L a_F^* = \left\{\frac{D_R}{D_L}\right\}^{-0.49} \times \left(34{,}500 \times v_g\right)^{0.86} \qquad \text{(A2)}$$

Where:

D_R = depth of the field reactor (m)
D_L = depth of the lab stripping column (0.815 m)
v_g = superficial gas velocity of the field reactor ($m^3.m^{-2}.s^{-1}$)

Using Equation 9, the N_2O volumetric mass transfer coefficient was calculated for all aerated reactor zones in the seven WWTPs. To determine the uncertainty associated with these field k_La values, two independent mass transfer correlation techniques were applied to the field data. The first correlation technique (Khudenko and Shpirt, 1986) empirically related the clean water oxygen mass transfer coefficient ($k_L a_{O_2}$) to v_g, reactor geometry and aeration bubble diameter, D_B.

The N_2O mass transfer coefficient ($k_L a_{N_2O}$) was then estimated in accordance with Higbie's penetration model (Capela et al., 2001):

$$k_L a_{N_2O} = k_L a_{O_2} \times \sqrt{\frac{D_{F,N_2O}}{D_{F,O_2}}} \qquad \text{(A3)}$$

Where:

$D_{F,N2O}$ = Molecular diffusivity of N_2O in water
= 1.84×10^{-9} $m^2.s^{-1}$ at 20°C (Tamimi *et al.*, 1994)
$D_{F,O}$ = Molecular diffusivity of oxygen in water
= 1.98×10^{-9} $m^2.s^{-1}$ at 20°C (Ferrell and Himmelblau, 2002)

The second correlation (Dudley, 1995) empirically related the clean water N_2O k_La to v_G, $D_{F,N2O}$, D_B and viscosity. Both correlations were also corrected for temperature, using a standard θ factor of 1.024 (Tchobanoglous et al., 2003):

$$k_L a_{T^oC} = k_L a_{20^oC} \times \theta^{T-20} \qquad \text{(A4)}$$

Both correlations are sensitive to bubble diameter, which was unknown for all WWTPs. Furthermore, the clean water k_La values must also be corrected for mixed liquor conditions and diffuser fouling (i.e. αF factor), which was also unknown. Therefore, a range of likely values was evaluated for both parameters: DB = 2–4mm (Hasanen et al., 2006) and $\alpha F = 0.4–0.9$ (Tchobanoglous et al., 2003). The results of these sensitivity analyses are shown in Figure A2.6.1B for the two correlations. It can be seen that the k_La model proposed for this study (Equation 9) fits well within the band of likely values and thus provides an independent validation of the adopted approach. The correlations also provide an uncertainty range for the k_La value used in Equation 5 (section 5.1.3). Using the full range of sensitivity analysis values calculated from the two correlations, a 95% confidence interval (t-dist, $\alpha = 0.05$) of k_La values was calculated (i.e. shaded area in Figure A2.6.1B). This confidence interval was then applied to each k_La determined by Equation A2. From Figure A2.6.1B, it can be seen that the values from Equation A2 sit in the lower quartile of the confidence interval generated from the literature correlations.

The mass transfer coefficients for quiescent reactor zones (i.e. primary sedimentation tanks, anaerobic zones, anoxic zones, secondary sedimentation tanks) were estimated using the empirical correlation technique of Van't Riet (1979). This relates k_La to the volumetric power input (P/V) for mixing. Similar to the approach for aerated zones, a range of likely P/V values was surveyed (2–8 $W.m^{-3}$) and used to construct a 95% confidence interval for each field reactor zone. However, the k_La values in these quiescent zones (indicatively 3–4 d^{-1}) were an order of magnitude smaller than those in the aerated zones. Therefore, the results are not especially sensitive to these values.

Appendix 2.7
WWTPs France

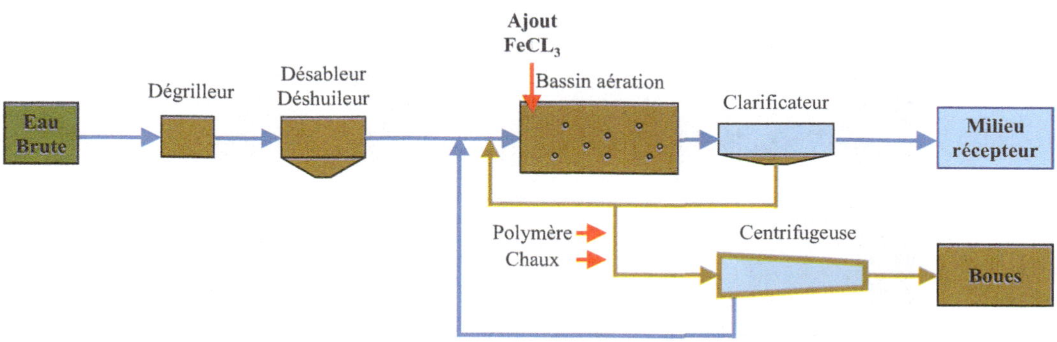

Figure A2.7.1 Scheme of WWTP 1.

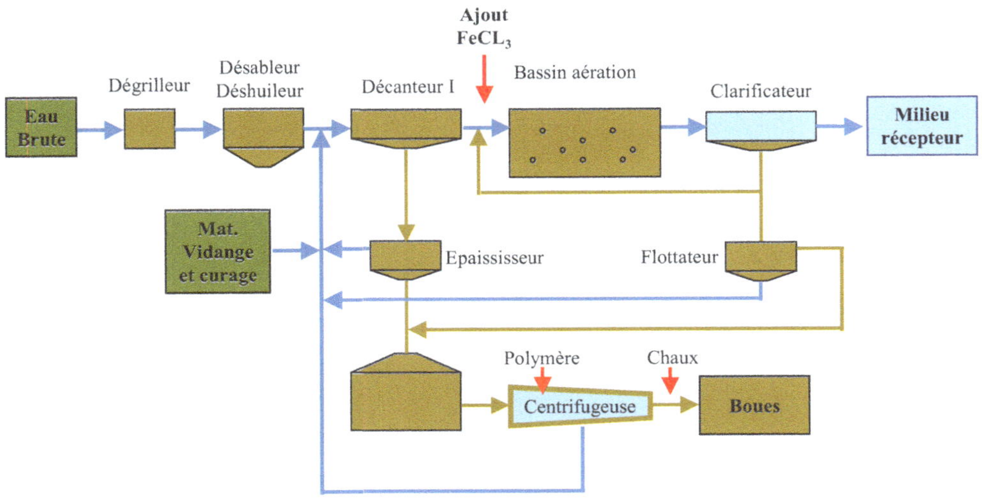

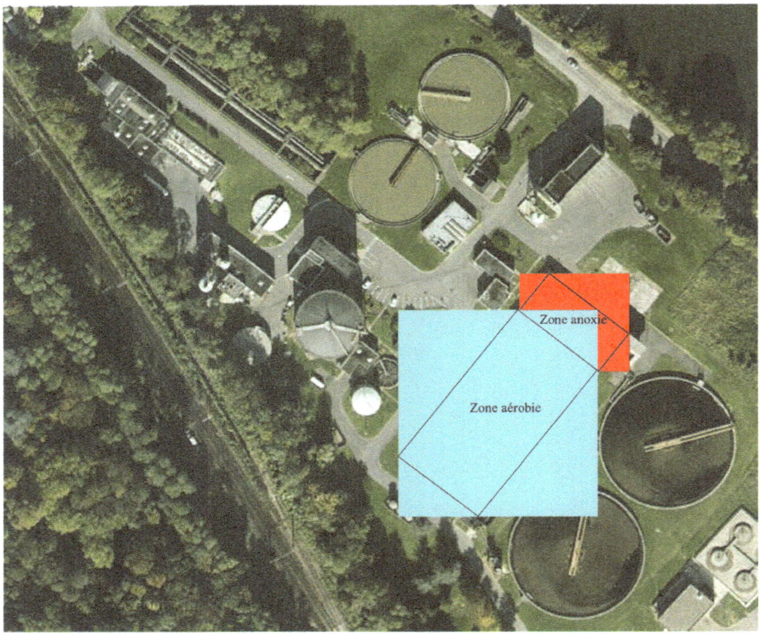

Figure A2.7.2 Scheme and photo of WWTP 2.

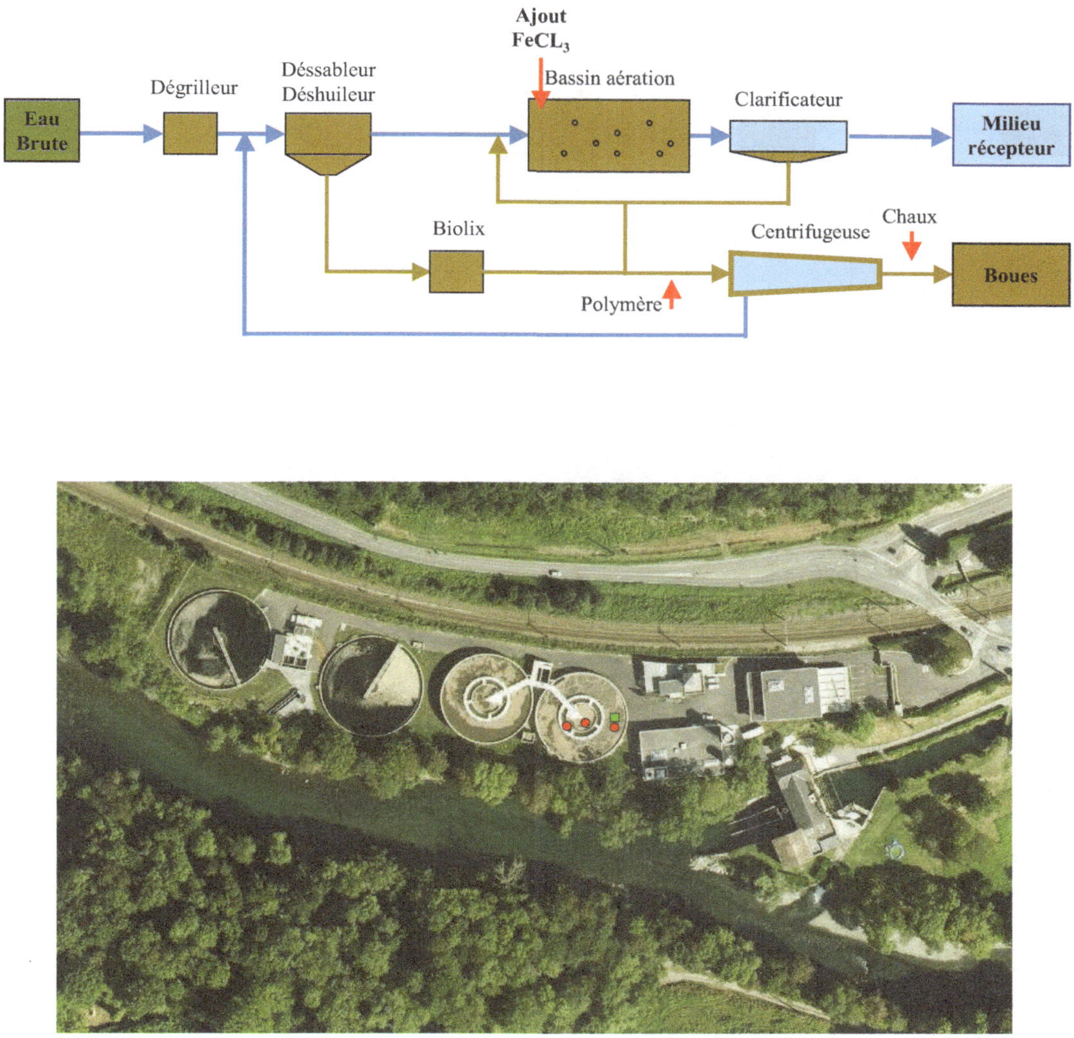

Figure A2.7.3 Scheme and photo of WWTP 3.

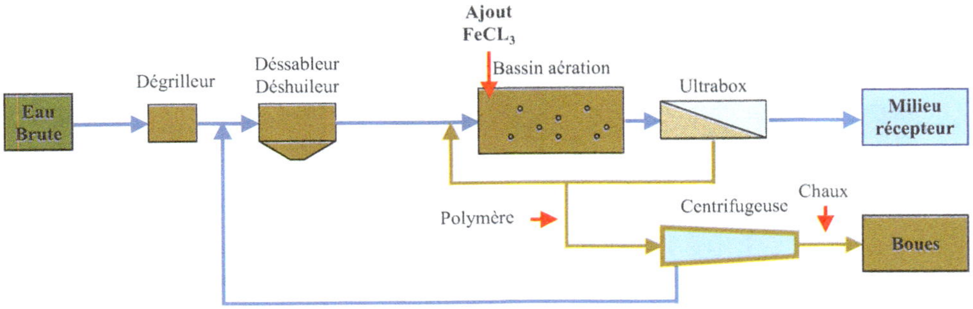

Figure A2.7.4 Scheme and photo of WWTP 4.

Appendix 2.8

DeKalb County's

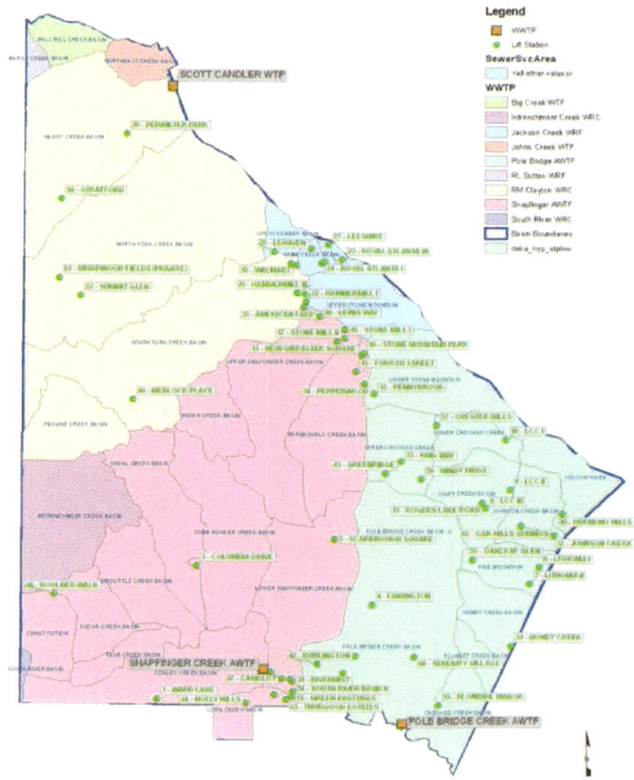

Figure A2.8.1 Map of DeKalb County's collection system.

IWA Publishing's authorised EU representative for General Product Safety Regulations is Diane D'Arras, 15 rue Duret, 75116 Paris, France, e-mail: safety@iwap.co.uk.

Printed and bound by CPI Group (UK) Ltd, Croydon, CR0 4YY
16/06/2026
02140376-0001